Communities and Forest Stewardship

REGIONAL TRANSITIONS IN SOUTHEAST ASIA

Mark Poffenberger

Rowena Soriaga

Peter Walpole

Asia Forest NETWORK

and

CFI

COMMUNITY FORESTRY INTERNATIONAL

2006

Foreword

The tropical forests that once cloaked much of Southeast Asia are largely gone. For thousands of years, the abundant rainfall in much of the region's mainland and island archipelago supported lush vegetation. Diverse human cultures have co-existed with these forests, evolving along the coastlines, lowland plains, river valleys and upland hills and mountains for millennia. Each group has developed unique sets of knowledge regarding the local ecology, the uses of hundreds of plants and animals, as well as ways to manipulate the forest to optimize their availability that has been passed down for generations. Over the past several hundred years, these human-forest relationships have changed at an accelerating pace, especially in recent decades. Expanding national populations, the demands of global markets, changing technologies, and ever improved access have opened the region's forests to rapid exploitation.

Southeast Asia's primary rain forests shrank from over 250 million hectares in 1900 to less than 60 million hectares in 1989.[1] Throughout this period, the vast majority of the region's forest land has been under the legal stewardship of national forestry agencies that have attempted, and failed, to manage over a hundred million hectares of industrial timber concession on a sustained yield basis. Other forest lands, under the authority of forest departments, have been cleared, either by large concessionaires, migrant farmers, or local residents, for agricultural use. While some forest conversion has been an inevitable result of population growth and development, large areas have been unsustainably exploited and have been left in degraded condition and now provide few benefits either in terms of environmental services, or economic productivity.

There is an urgent need for a fundamental forest paradigm shift in Southeast Asia that will transform the ways in which governments perceive and manage their forest environments. For over a century, the dominant model has dictated that professional foresters should hold primary control over the majority of each nation's land area, typically ranging from 30 to 70 percent, under the ultimate direction of the central government. This model has failed dramatically. Further, attempts for legislative and policy reform have been tentative in most countries, and while the industrial extraction concept has been generally discredited, no alternative approach has yet received a strong endorsement. This book explores the process of change that is currently under way in the forestry sector in Southeast Asia, and attempts to identify where progress is being made in conceptualizing and operationalizing participatory modes of stewardship, and where barriers are constraining change.

As this book reports, there is growing interest in exploring ways to link the formal forestry sector to informal and emerging initiatives of rural communities. This is a challenging task as the forestry sector in Southeast Asia, as in many parts of the world, continues to be characterized by a technocratic, hierarchical, and bureaucratic orientation to management, and a need to retain control and authority. The informal forest cultures of the region, by contrast, emphasize a socio-cultural orientation and a multiplicity of use. As one scholar writes in contrasting the formal forestry paradigm with indigenous forest culture:

> *It concentrates on a definition of forest that is uniquely physiognomic and silvigenetic. This represen-tation denies the reality of millennia of co-evolution between human societies and forests, and it sends the economic, social, symbolic and political dimensions of forest history back to a secondary position.*[2]

The authors of this book take the position that the most promising approaches to restoring and conserving Southeast Asia's rapidly disappearing forests will be through the engagement of rural communities in their management. This strategy does not represent a "magic bullet" that can quickly resolve the forest crisis in Asia. Now under way for twenty years or more in some countries, steps towards meaningful engagement of forest-dependent peoples in resource stewardship partnerships is still in an early stage of implementation. Nonetheless, it represents a sea of change from earlier in the twentieth century when forest-dependent people were stigmatized, and viewed as "encroachers," "illegal inhabitants," "backward tribals," "practitioners of slash and burn," and "criminal castes." The long term transition in the forest sector is something that will likely continue to unfold throughout the 21st century. Community capacities lie in their commitment to, and knowledge of place, their cultural ties to the forest, and their economic dependence on the forest, both for subsistence goods, cash, and environmental services. There remains, in many parts of Southeast Asia, considerable social capital reflected in the institutions, values, and practices of indigenous societies. On balance, rural communities are under pervasive and growing pressures from urbanization, privatization, modernization, commercialization, and individualization. All these forces have the capacity to erode indigenous culture and institutions that have played an important role in mediating community use of forests and natural resources in the past.

Recognizing the role communities can play in the future of forest management in Southeast Asia, there has been a growing movement in the region to formulate laws, policies, and programs to formally involve rural people forest stewardship.

In the face of rapid deforestation, and the resulting loss of upland bio-diversity, torrential downstream floods, and disruptive urban brown-outs, Southeast Asian governments, city dwellers, and rural communities have grown increasingly concerned over the deteriorating state of their forests and watersheds. National media are widely documenting and disseminating information through TV, newspapers, and radio regarding continuing forest destruction. Urban people and villagers across the region are increasingly aware of the need for forest conservation and more sustainable use. This concern is reflected in recent laws and policies to protect the environment and involve communities in management.

Throughout the 1990s, many Southeast Asian nations have been actively engaged in exploring innovative approaches to community-based resource management, attempting to integrate traditional resource stewardship practices into modern governance structures. This process of devolving management rights for public forest lands to local populations is being supported through a variety of policy initiatives and legal instruments including decentralization acts and local governance ordinances, as well as new forest and environmental laws. Further, a number of governments have formulated specific community forestry sub-decrees, government orders, and guidelines to facilitate the transfer of stewardship authority to local groups. Many international development agencies consider community-based natural resource management a priority component of their assistance strategies.

In 2001, with support from the European Commission's Tropical Forests Program and the United States Agency for International Development's East Asia and the Pacific Environmental Initiative, the Asia Forest Network and Community Forestry International initiated the Community Forest Management Support Project (CFMSP) for Southeast Asia to facilitate forest sector transitions under way in the region. The project was designed to respond to needs at the community, national, and regional level through a variety of interventions. At the regional level, CFMSP organized a series of workshops and cross visits to stimulate exchange between countries engaged in developing community forest management policies and programs. At the national level, CFMSP provided financial and technical assistance to country working groups, NGO networks, and donor dialogues that were developing policy frameworks and national strategies to encourage forestry sector transitions that engaged communities as principal partners. At the field level, CFMSP worked with partner organizations implementing community forestry initiatives, by providing small grants, technical assistance, and support with documentation.

One component of CFMSP was to collaborate with field project partners to produce one case study from each of the five participating Southeast Asian countries, including Cambodia, Indonesia, the Philippines, Thailand, and Vietnam. The creation of resource management partnerships linking communities and local governments is a strong theme in each of the five case studies. So too is the process of building community abilities and confidence to protect and regulate access to their natural resources. The case studies primarily examine changes occurring in the past five years. For the most part, the progress made in stabilizing local resources, building community institutions, resolving conflict with local government and neighboring villages, and in establishing a sustainable system of management has been dramatic. These experiences from five corners of Asia indicate that the trust planners, NGOs, development agencies, and the larger civil society is gradually placing in region's rural villagers is not misplaced. At the same time, as is apparent from each of the cases, the need for financial, technical, and political support is vast. A great deal of damage has been done to the region's forests in recent decades due to national policy, as well as field-level management failures. An equally extensive effort will be needed to restore these critical ecosystems and community relationships with them. The case studies suggest that a long term investment in building the capacity of communities and local governments to sustainably manage much of Southeast Asia's forests would be a strategic one.

This book focuses on some of the changes that have taken place in the forest sector between the early 1980s and 2005. As such, it chronicles some of the early legislative, policy, and programmatic actions that have been taken to devolve authority and enable community forest management and the prospects they offer for future change.

Dr. Mark Poffenberger

Executive Director
Community Forestry International

1 Scott, Margaret. 1989. "The Disappearing Forests" in *Far East Economic Review* (January 12, 1989), p. 32.

2 Michon, G. 2005. "Domesticating Forests: How Farmers Manage Forest Resources" (Bogor: FORRESASIA), p. 4.

Acknowledgements

The authors would like to acknowledge the contributions of many individuals who have helped inform this book. Many of these resource persons are leading community forestry specialists in their own country and are contributing to national forest sector transitions. The contributors are drawn from national forestry agencies, international research and development organizations, universities and NGOs. Much of the information presented in the book was contributed during regional meetings and field workshops convened under the Community Forest Management Support Project (CFMSP) that brought specialists from Southeast Asian nations together to share experiences with community forestry between 2001 and 2004. We are extremely grateful for their input. Contributors, editors, and commentators involved in the book included the following individuals, as well as hundreds of rural people in the field areas.

Special thanks are due to our CFMSP country partners who have substantively informed the contents of this book. In Cambodia they include, Ty Sokhun, Chea Sam Ang, Ly Chou Beang, Sokh Heng, Lao Sethaphal, Patrick. T. Evans, Prak Marina, Melissa Marschke, Rob Oberndorf, Kiran Paudyal, Graeme Brown, Amanda Bradley, Yang Phirom, and Todd Sigaty. In Indonesia, our colleagues included N. Juni Adi, Muhamad Chehafudin, Faisal Fuad, Krustanto, Soetrisno, Iman Santoso, and Suhardijono. In the Philippines, we are grateful to Romeo Acosta, May Blanco, Sylvia Miclat, Jojo Parreño, Victor Ramos, Rene de Rueda, and Dionisio Tolentino for their contributions. In Thailand, we appreciate the continuing support of Ruang Janmahasatien, Jessada Kaewchote, Chaleo Kanjan, Korkiat Kaysornsiri, Pearmsak Makarabhirom, Komon Pragtong, and Udhai Thongmee. And in Vietnam, CFMSP has benefited from the long term collaboration of Nguyen Huy Phon, Nguyen Huy Dzung, Ho Manh Tuong, and Nguyen Hai Nam.

Special thanks are due to Jon Lindsay at FAO Rome and Don Gilmour, for their careful review of the manuscript and thoughtful comments. The authors would also like to thank Mark Sandiford at UNDP/ SGP-PTF, and Cor Veer for their contributions to the program. We also acknowledge the stimulating discussions with colleagues in India including: S. Palit, Urmila Pingle, Emmanuel D'Silva, Peter Jipp, Parmesh Shah, and Sanjay Upadhayay.

We would like to acknowledge the contributions by the staff of the Asia Forest Network and the Environmental Science for Social Change including Arlen Salgados-Cañares, Noelyn Dano, Akiko Inoguchi, Dennis Maliwanag, Steve Magallano, Abigail Mejorada, Liezl Samonte-Bunda, Fidel Supremo and Gussy Villa-Real of AFN and Efrem Cimafranca and Mariel de Jesus of ESSC. Finally, we are indebted to Kate Smith-Hanssen, Ph.D., CFI, for her editorial assistance and guidance in the development of the final manuscript.

Asia Forest Network and Community Forestry International would like to thank the European Commission and United States Agency for International Development for their support. We would also like to express our appreciation to our partner organizations who are engaged in implementing a new generation of community forestry laws, policies, and programs. Finally, we would want to emphasize the tremendous effort being made by thousands of rural communities across Southeast Asia that contribute to forest protection, conservation, and the sustainable management of the planets natural ecosystems. They require the support of national governments and the international community.

Table of Contents

List of Tables

List of Figures

List of Boxes

List of Maps

Annex

List of Acronyms

ARP	–	Agricultural Rehabilitation Program
AFN	–	Asia Forest Network
AIT&C	–	Asia Information, Technology and Communications Programme
APFC	–	Asia Pacific Forestry Commission
ADB	–	Asian Development Bank
BFL	–	Basic Forestry Law
CGFP	–	Cambodian German Forestry Project
CWWG	–	Carood Watershed Working Group
CIFOR	–	Centre for International Forestry Research
CADT	–	Certificate of Ancestral Domain Titles
CALC	–	Certificate of Ancestral Land Claim
CCFS	–	Certificate of Community Forest Stewardship
CS	–	Certificate of Stewardship
CFA	–	CF Agreement
CFIP	–	Chamber of Furniture Industries of the Philippines
CADC	–	Claims of Ancestral Domain Certificates
CEP	–	Coastal Environment Program
CFM	–	Collaborative Forest Management or Co-Management
CPR	–	Common Property Resources
CTF	–	Communal Tree Farming
CL	–	Commune Law
FKKM	–	Communications Forum for Community Forestry-Forum Komunikasi Kehutanan Masyarakat
CFM	–	Community Forest Management
CFMA	–	Community Forest Management Agreement
CFMSP	–	Community Forest Management Support Program
CFSD	–	Community Forest Sub-Decree
CF	–	Community Forestry
CFAC	–	Community Forestry Alliance for Cambodia
CFI	–	Community Forestry International
CFP	–	Community Forestry Program
CRMF	–	Community Resource Management Framework
CBFM	–	Community-Based Forest Management
CBFRM	–	Community-Based Forest Resource Management
CBNRM	–	Community-Based Natural Resource Management

CONCERN – Concern Worldwide
C – Constitution
CBD – Convention on Biological Diversity
DARD – Department of Agriculture and Rural Development
DFWR – Department of Forest and Water Resources
DFW – Department of Forestry and Wildlife
DENR – Department of Environment and Natural Resources
GTZ – Deutsche Gesellschaft fur Technische Zusammerarbeit
DPRD – Dewan Perwakilan Rakyat Daerah; District Legislative Assembly
EAPEI – East Asia and the Pacific Environmental Initiative
EP – Environmental Protection Law
ESSC – Environmental Science for Social Change
EC – European Commission
EU – European Union
FAR – Family Approach to Reforestation
FAO – Food and Agriculture Organization
FA – Forest Administration
FCM – Forest Concession Management
FDA – Forest Development Agency
FIO – Forest Industry Organization
FIPI – Forest Inventory and Planning Institute
FLMP – Forest Land Management Program
FLMA – Forest Lease Management Agreement
FREC – Forest Resources and Environment Center
FSSP – Forest Sector Support Program
FL – Forestry Law
FSP – Forestry Sector Project
IPs – Indigenous Peoples
IPRA – Indigenous Peoples Rights Act
WALHI – Indonesian Forum on the Environment-Wahana Linkgungan Hidup Indonesia
APKINDO – Indonesian Wood Panel Association
IFMA – Industrial Forest Management Agreement
IRMP – Integrated Rainforest Management Project
ISF – Integrated Social Forestry
IPF/IFF – Intergovernmental Panel on Forests/Intergovernmental Forum on Forests
ICRAF – International Centre for Agroforestry Research
JFM – Joint Forest Management

LL	–	Land Law
LFM	–	Local Forest Management
LGU	–	Local Government Unit
LIUCP	–	Low Income Upland Communities Project
MARD	–	Ministry of Agriculture and Rural Development
MAFF	–	Ministry of Agriculture, Forestry and Fisheries
MoF	–	Ministry of Forestry
MoFEC	–	Ministry of Forestry and Estate Crops
MNRE	–	Ministry of Natural Resources and Environment
NCIP	–	National Commission on Indigenous Peoples
NFP	–	National Forestation Program
NIPAS	–	National Integrated Protected Area System Act
NRMP	–	Natural Resource Management Program
NRM	–	Natural Resources Management
SKEPHI	–	NGO Network for Forest Conservation In Indonesia-Sekretariat Kerjasama Pelestarian Hutan Indonesia
NGO	–	Non Government Organization
NGA	–	Non-Government Agencies
NTFPs	–	Non-Timber Forest Products
NDF	–	Northern Development Foundation
NFN	–	Northern Farmer's Network
OXFAM	–	Oxfam G.B.
PFM	–	Participatory Forest Management
PF	–	Participatory Forestry
PLUP	–	Participatory Land Use-Planning
PRA	–	Participatory Rural Appraisal
PO	–	People's Organization
PP	–	Peraturan Pemerintah
PWG	–	Philippine Working Group
CD	–	Prakas on Cardamom Mountains
PAMB	–	Protected Area Management Board
PAFO	–	Provincial Agriculture and Forestry Office
RECOFTC	–	Regional Community Forestry Training Centre
RRMP	–	Regional Resources Management Project
RFD	–	Royal Forest Department
SM-HDP	–	Sam Mun Highland Development Project
SALT	–	sloping agriculture land technology

SGP-PTF	–	Small Grants Programme to Promote Tropical Forests
HPHH	–	Small-scale Timber Harvesting
SFC	–	State Forest Corporation
SFM	–	Sustainable Forest Management
SMRP	–	Sustainable Management of Natural Resources
TAO Act	–	Tambon Administration Act
TAOs	–	Tambon Administrative Organizations
TLAs	–	Timber License Agreements
UMA	–	Ubay-Mabini-Alicia
DFID	–	UK Department for International Development
UU	–	Undang-undang
UN	–	United Nations
UNFF	–	United Nations Forum on Forests
USAID	–	United States Agency for International Development
UDP	–	Upland Development Project
VFI	–	Village Focus International
ARuPA	–	Volunteers' Alliance for Saving Nature
WCMO	–	Watershed Conservation and Management
WMU	–	Watershed Management Unit
WFF	–	Wonosobo Forest Forum
WB	–	World Bank
WRI	–	World Resources Institute
WSSD	–	World Summit on Sustainable Development
WWF	–	World Wildlife Fund for Nature

PART I

Communities and Forests

Community forest management activities date back thousands of years, long before the advent of agriculture. The felling of trees for fuelwood, the collection of fruit and leaves, the hunting of game, and the use of fire to clear undergrowth and stimulate grass production for wild animals were likely routine activities for forest dwellers during the Neolithic period and were probably practiced at least 50,000 years ago. Archaeological findings from numerous sites in Southeast Asia indicate that forest and cave-dwelling communities hunted widely in the tropical forests of Southeast Asia from the coast of Vietnam, the Malay Peninsula, and Sumatra.[1] The clearing of forest plots to plant seeds, a form of early swidden farming, may well predate agriculture in the flood plains and deltas. The annual setting of ground fires by indigenous peoples dramatically shaped the vegetative landscape in many parts of the world.

> *The most powerful, effective, and widely employed tool in the native repertoire for directly manipulating the environment was undoubtedly fire. Indigenous groups used fire for a variety of purposes, including stimulating new plant growth and inducting early stages of succession; creating and sustaining vegetational mosaics with numerous ecotones beneficial to animal life, controlling plant disease and insect infestation; and increasing the frequency and range of useful plant species; eliciting desirable plant growth characteristics minimizing the severity and number of uncontrolled wildfires; and facilitating hunting by the reduction of undergrowth.*[2]

Over the centuries, forest management grew in sophistication and took a multiplicity of forms, both in the institutions that guided it, and in the technologies that were utilized. As populations expanded, customary laws were developed to govern forest usufruct and provide tenure arrangements supporting the values and traditions of the society and the production systems that its people employed. Many cultures held forests under collective stewardship, viewing their role as custodians of resources kept in trust for future generations and shared by many other species. Authority structures were diverse including hereditary chieftainships, tribal councils, groups of elders, some more hierarchical and autocratic, and some more democratic.

This is not to imply that community management systems always lead to a harmonious balance between human use and ecological stability. As William Cronon notes:

> *If we assume a priori that cultures are systems which tend towards ecological stability, we may overlook the evidence from many cultures—even pre-industrial ones—that human groups often have significantly unstable interactions with their environment.*[3]

For example, Jared Diamond, in his book *Collapse: How Societies Choose to Fall or Succeed,* argues persuasively that the civilization on Easter Island disappeared after inhabitants deforested the isolated island sometime between 900 and 1500 AD.[4] Japan also ran into serious resource problems after felling much of its forests from 800 to 1700 AD, only to develop excellent forest management systems over the past three centuries.

It is clear, then, that learning how to sustainably manage natural resources has been a process of trial and error for communities for thousands of years.

This caveat aside, it is evident that pre-industrial societies, especially forest-dependent cultures, have been in an age-old and ongoing dialogue between the local communities and their environment, and that it has often been informed by a substantial knowledge of place, based on long experience, that frequently resulted in reasonably sustainable patterns of use. Local knowledge was reinforced by a motivation to sustain the natural resource base, a sort of place-based, inter-generational accountability. While community resource management did not produce a static environment, many indigenous cultures with a high level of forest-dependence have supported a dynamic equilibrium within their ecosystems, where fire, felling, gathering and planting facilitated changes that enhanced diversity and regeneration. To that end, William Cronon comments, "the best measure of a culture's ecological stability may well be how successfully its environmental changes maintain its ability to reproduce itself."[5]

Over the past two centuries, the nationalization of much of the world's forests has generally eroded the authority of indigenous community-based forest managers and their institutions. Forest departments have frequently been vested with jurisdiction over the "state domain," yet with limited fiscal and human resources these agencies have faced growing difficulties protecting millions of hectares of remote forest lands. In many developing nations, especially post-World War II, much of the public forest estate has fallen under the control of politicians and private sector interests. Forestry agencies have often been co-opted by such influential actors, facilitating unsustainable industrial logging operations that have led to extensive forest loss, social displacement, and environmental degradation, while contributing little to national development. In many developing nations, civil society has had little input into state land management decision-making; however, this has started to change in the last decades of the twentieth century as public concern over forest degradation has increased and the political leadership in a growing number of countries has begun to respond.

Community Forestry Systems

Indigenous Community Forestry

Indigenous community forestry systems refer to traditional resource use practices of communities

Indigenous Community Forestry Systems in Vietnam. A Nung woman heads home with bundles of fodder. The responsibility of forest management lies in the hands of the community, whose proximity to the resource provides for effective resource management.

that have a long history of residence and forest use in an area. The phrase "indigenous peoples" has no globally accepted definition, but the term has gained popularity in international dialogues concerning "ethnic minorities," "tribal populations," "first nations," and "native peoples."[6] Many indigenous peoples hold their resources collectively and, among forest-dependent peoples, practice swidden cultivation. Indigenous community forestry systems are often guided by traditional institutions that oversee forest use including the allocation and clearing of land for swidden fields, the protection of regenerating forests, water sites, and sacred groves, the felling of trees for house building, and the resolution of conflicts.

Indigenous people face discrimination and are often socially and economically marginalized since they are usually a minority population within their own countries. Since communal systems of land stewardship frequently receive little recognition under the agrarian laws of many Asian nations, indigenous peoples are under increasing pressure to privatize their collective land and forest resources, a process that can erode their social cohesion, traditional values, and cultural identity. Privatization also weakens indigenous resource management institutions, leading to a breakdown of local authority structures and forest use control mechanisms that has contributed to the acceleration of deforestation in Asia. In response to this problem, some Asian nations have passed legislation recognizing ancestral domain claims and communal tenure, although only in the Philippines have such policies been widely implemented.

Indigenous community forestry systems are under growing pressure, yet they represent some of only a few proven systems for successfully conserving and managing millions of hectares of upland forest in insular and mainland Southeast Asia. But, because they are generally not formally recognized by government, they remain the hidden face of forest management and continue to be held under the nominal authority of state forest agencies. Due to the legitimate claims of indigenous peoples for their forest lands, legal and policy issues go far beyond only forest management but include land issues, human rights, social equity, and livelihood concerns.

Responsive Community Forestry

Unlike indigenous systems, responsive community forestry systems are not based on a long history of use or ancestral claims to land and resources. More commonly, responsive systems emerge among rural communities whose shared dependence on forest resources drives them to establish regulations governing use arising from pressures on, or threats to, the forest and related resources. For example, a farming community may find that forest degradation is creating water shortages and that watershed protection is needed to restore hydrological function. Alternatively, a migrant community may find that a growing population is creating fuelwood scarcities, and that greater forest protection and managed harvesting and distribution of firewood is needed to ensure equitable and sustainable use.

While forest management systems have been developing for thousands of years, anecdotal evidence suggests growing demographic and industrial pressures in recent decades have forced many rural communities to invest more time in proactively managing forest lands by modernizing and formalizing management organizations, documenting and codifying rules of use, and placing greater restrictions on access. On a continuum from completely open access systems to more tightly controlled community management systems, there appears to be a worldwide trend for more thoughtful stewardship. This is apparent in many rural areas of Southeast Asia, even where external programs and projects are not present. In examining the community forestry case studies presented in Part IV, the reader will find that timelines of events reflect an initial concern among villagers regarding a negative environmental change followed by community-based activities to address the problem. Imposition of community-based controls to moderate forest use or facilitate ecological recovery are an increasingly common phenomenon in Southeast Asia, especially in areas where forests face external threats or environmental services functions of forests are declining.

While some indigenous groups are struggling to establish their legitimacy in the absence of government support, other rural communities are exploring ways to enhance their capacity to regulate forest use through the formalization of institutions, codification and elaboration of rules and regulations, and the development of formal management plans. Spontaneous community forestry groups emerging in Southeast Asia take diverse forms. In some cases, spontaneous community forestry groups are being identified and brought within the framework of national community forestry programs and are beginning to receive external recognition and sponsorship; however, many spontaneous groups remain informal bodies and represent another hidden side of the unknown revolution taking place in Southeast Asia's forests.

Sponsored Community Forestry

Sponsored community forestry systems refer to community groups that have been formed by and supported through external agencies, whether they are government, forest departments, NGOs, or other institutions. These systems benefit from the technical and financial support provided by the sponsoring organization, but they do not have the same kind of community ownership as the indigenous and spontaneously formed groups. From that standpoint, they have less chance of being sustained beyond the project period and are often less effective in implementing forest management activities and plans. They may also be more susceptible to manipulation by external actors. Increasingly, as national community forestry support programs have developed, indigenous and spontaneous groups have begun receiving external support. This approach allows for the community-based institutions that have evolved independently to benefit from external recognition and support. As a consequence, it is apparent that many community forestry groups may reflect a blend of indigenous, spontaneous, and sponsored elements in their experience.

Community Forestry Terms

Terms commonly used to describe community involvement in forest management include collaborative forest management, social forestry, community forestry, joint forest management and

Responsive Community Forestry Systems in Java. While forest management systems have been developing for thousands of years, anecdotal evidence suggests growing demographic and industrial pressures in recent decades have forced many rural communities to invest more time in proactively managing forest lands.

others. Many of these terms have evolved from development agency and national government programs to support community engagement in state forest land management, or local forest production initiatives. It should be noted that there is a good deal of overlap in the meaning of these terms, and many are used interchangeably. There are also strong national preferences. These terms as well as others are briefly defined as follows:[7]

Agroforestry

Agroforestry covers a wide variety of poly culture systems that include trees, shrubs, perennial crops, annual crops, herbs, and climbers. Some agroforestry systems include over one hundred species of plants, while others are far less diverse. By definition, agroforestry systems will be multi-storied with a canopy and understory. While most agroforestry systems are managed by households, they may also be coordinated through community organizations and practiced on communal land. Agroforestry in most cases is seen as an acceptable and recognized element of many community forms of forest management.

Ancestral Domain Management

Ancestral domain, or traditionally held land, is informally acknowledged in all Southeast Asian countries on the basis of ancestral use of natural resources and the resident cultural practices. The Philippines has legislated a formal system of recognizing such areas including the development of a management plan and more recently, the granting of titles and ancestral domain certificates.

Collaborative Forest Management or Co-Management (CFM)

Collaborative Forest Management or Co-Management refers to a partnership by which stakeholders agree on sharing management duties, rights, and responsibilities for an area of forest. The stakeholders typically involve the agency in charge of the resource, and the various associations of local residents, local and traditional authorities, industries, business, research institutions and others.

Community-Based Forest Management (CBFM)

In the Philippines, CBFM is widely used to refer to forest management by or with the local community. CBFM includes traditional forms of forest management, more recently self-mobilized community forest initiatives, commune or municipal forests, shared-ownership with state, and forms of collaborative management between state and community organizations.

Community Forestry (CF) or Community Forest Management (CFM)

These are general terms that include indigenous forest management systems, spontaneous forest management systems, as well as programs initiated by government. Used in Nepal from the early 1970s, forest groups apply to manage local forests and to utilize the products (on a shared arrangement) with government. CF and CFM are now widely used to denote many forms of people-based forest management.

Farm Forestry

Farm forestry generally refers to tree-planting on private land by farmers who often form community groups in order to pursue joint marketing, processing, and mutual learning.

Joint Forest Management (JFM)

JFM developed in India to refer to community involvement in managing forest reserves by formalizing the sharing of products and responsibilities between local communities, forest department, and other user groups. JFM agreements are based on a contract specifying the distribution of authority, responsibility, and benefit between villagers and state forest services.

Local Forest Management (LFM)

LFM refers to the actions of people living near a forest to maintain or enhance the forest and improve their well being. LFM is similar to spontaneous community forest management where local people enhance the sustainability of forests, acquire a share of benefits, maintain control over decisions related to resources, and competing demands are resolved in ways that reduce conflict and enable synergies. This term implies that management can be mainly done locally with minimal interference from outside. The term is also used to refer to central governments that devolve forest management to local individuals and institutions, both within and outside government.

Participatory Forestry (PF) or Participatory Forest Management (PFM)

PFM is an umbrella term that could include all the above terms and forms.

Public Participation in Forestry

This is a very broad term that includes the many ways people can influence forestry, but not necessarily in shared (collaborative) decision-making.

Social Forestry

Social Forestry was used by the Indian government in the 1970s to refer to tree planting projects, usually monoculture, that was done on village land, not on forest reserves. The term is now used more generally to refer to forest management with social goals for local people.

Emerging Change

The formal emergence of community-based resource management paradigms and their support by national governments and donor agencies in Southeast Asia has largely taken place during the past twenty years. David Korten, writing in the 1980s and an early proponent of this approach, based his support on the limitations of public administration systems throughout the region.

> *Public development efforts of the past few decades have seen increasing extension of state authority throughout Asia into affairs once the preserve of local custom and control. In many respects this is a natural and necessary aspect of the modernization process, drawing local communities into participation in larger national and global systems. All too often however, in its enthusiasm for modernizing and rationalizing resource management, the state has underestimated the extent and capacity of the systems by which people*

Mount Diwata in Agusan del Sur, Philippines. A co-management agreement has been established between the government and the Department of Environment and Natural Resources (DENR) for the management of this critical forested area that is the source of water for the municipality of San Francisco.

> *have learned through long and often difficult experience to manage locally available resources to meet their own self-defined needs. At the same time the state has often seriously over-estimated its own ability to manage these same resources.*[8]

Even twenty years after Korten wrote this assessment, government capacity to sustainably manage hundreds of millions of hectares of forest land remains an unfilled dream. In many parts of Southeast Asia, communities remain the keepers of forests on a day-to-day basis. Forest departments, often hundreds or thousands of kilometers away, hold responsibilities for management in name alone. Yet, the illusion of state management remains the dominant image on the national political landscape.

Retaining legal authority over forests, even if ground-level management capacity remains extremely limited, has been advantageous for national governments and their political leadership as it provides a framework to allocate timber and land use rights to private sector interests. Many forest agencies are easily manipulated by political elite, allowing resource privileges to be allocated and key political coalitions to be maintained.[9] Yet the failure of the state management paradigm to sustain the region's forests has led to growing pressure at both the international and national level for reforms in resource stewardship systems and has forced a gradually growing recognition of the important roles that communities play in sustaining upland watersheds, forests, coastal reefs and other resources. The pressures are not only from above but from below, where a groundswell of rural unrest is increasing as natural resources are usurped by the industrial and commercial sector while rural communities are left to deal with the environmental consequences.

Although community forest management has been slow to be formally recognized and national governments critiqued for a lack of commitment to share significant forest control benefits with some of the world's poorest communities,[10] the new laws, policies, and programs emerging in many Southeast Asian countries is a positive trend of devolving resource management rights and responsibilities across the region. Since nationalization of the world's forest lands is a process that has been under way for several centuries, shifting management direction from greater centralization of control to decentralization of resource management is a long-term process. This is particularly true, given the inherent resistance of power centers to transfer meaningful rights to other social entities, especially poor rural people.

In the mid-1980s, a group of Asian scholars and foresters contributed articles to an anthology entitled *Keepers of the Forests*. The book concluded that only through a convergence of state and local systems of management could stability re-emerge in the Asia's forests. As William J. Burch, then Director of the Tropical Resources Institute at Yale University wrote in the foreword:

> Keepers of the Forest *sings a tale that expresses how once upon a time there were traditional patterns of order and balance between people's needs and the sustainability of their resources. Authority and accountability were close to the source of need and nature. Then came a period of disorder and destruction as resources were redefined to meet centralized, commercial goals of distant accountability and whimsical market forces. Now, with the rise of modern Asia, there is the emergence of redemption and restoration as traditional strategies are rediscovered and adapted to today's realities. In Southeast Asia both the pressures of the population-land ratio and the emergence of new desires and resources are being enfolded into new ways of fulfilling older vision.*[11]

In many ways, *Communities and Forest Stewardship: Regional Transitions in Southeast Asia* is a sequel to *Keepers of the Forest*, assessing how change is slowly progressing since the original authors proposed their vision for transition and transformation across the forestry sector in Asia. In retrospect, early expectations for a smooth and systematic transition that would rapidly devolve greater forest management authority to rural communities were unrealistic. While progress toward national community forestry systems is apparent, headway towards widespread decentralization of natural resource management has been limited by low technical and administrative capacity at the community level, and resistance by government agencies as well as elite stakeholders.

Communities and Forest Stewardship: Regional Transitions in Southeast Asia charts the continuing evolution of emerging policy frameworks that enable the formal entry of communities as co-managers in the state forestry sector, examines the laws and regulations pushing decentralized governance and devolution of local resource management tasks to elected officials, and explores the ways legislation and programs have affected the lives of forest-dependent communities.

It is not a story of great success, dramatic changes, or a call to revive ancient natural resource management practices. However, it does highlight the continuing deterioration of the social fabric of many of Southeast Asia's indigenous societies as millions of migrants pour into once isolated forest tracts with little or no tradition of forest stewardship. And, it underscores the limited authority and resources for forest protection that communities currently enjoy. Until "authority and accountability" are re-established proximate to the resource, sustainable management will remain a distant goal. And, in many places only local communities and indigenous forest-dependent people are best positioned to take on this role.

While state control remains weak in much of rural and upland Southeast Asia, it still does not easily allow the transfer to or recognition of local rights. Legally and financially empowered by the central government, state representation in the form of technical forestry agencies or officials are often supported by local political and economic elite that create extensive networks of corruption that continue to marginalize communities and degrade the forests.

As Korten notes, "the natural dynamics of bureaucratic functioning create a substantial danger that programs promoted by government in the name of community management may instead contribute to processes of concentration and marginalization,

Children of the forest. Community forest management is not simply a question of forest management but tightly tied to human rights and social development concerns.

which work against the broader community interest."[12] As a consequence, he notes that structural, policy, and value changes are needed to reorient government if it is to be an effective community partner. Why then should government bother given it is a difficult solution?

He provides three arguments. First, given the immense social and environmental diversity found in most Asian nations, only local communities have the adaptive capacity to design management strategies for site specific contexts. Second, local communities have abundant resources including knowledge, and human resources that can be applied to management if tenure security and political support are present. And third, democratic societies assume that authority should rest with people who will bear responsibilities for their actions. Since forest-dependent communities are closest to the resources, they will be the first to bear the consequences of poor management and therefore are more likely to be responsible and accountable for their actions.[13]

In order for true partnerships to emerge between government and rural communities, significant "reversals in learning" must occur that allow outsiders, whether they be government officials, development workers, or NGOs to learn from the rural poor.[14] Attitudinal biases that favor modern technology over indigenous, local knowledge, men over women, and urban values over rural ones, all create barriers to integration and acceptance of many aspects of rural systems of resource use. The struggle for forest land and resources will continue well into the twenty-first century and beyond, for some kind of management is an ongoing requirement as long as human pressures are present. Community forest management does not represent the easiest path to sustainable use, but it may offer the best prospect for success.

Part I of *Communities and Forest Stewardship: Regional Transitions in Southeast Asia* briefly describes the historical, political and economic context that frames contemporary forest sector changes. Part II explores the historical and social context in which this drama is being enacted in Southeast Asia. Part III describes the nature of policies and laws that effect community control over forests, and Asia-wide trends that reflect the ways national societies are responding to public forest lands management. Part IV examines important shifts occurring in the forest policy sector in Cambodia, Indonesia, the Philippines, Thailand and Vietnam, following the decline of the industrial forest management paradigm over the last two decades and the emergence of a new generation of environmentally and socially-oriented policies and legislation. This section will explore how these new policies, laws, and national programs are affecting forest-dependent peoples across the region in an effort to track the transition in forest management on the ground. Part V identifies and discusses forces that will continue to promote and constrain broad-based transitions in forest management in Southeast Asia in the coming decades, and then examines the broad range of human-ecological contexts in which CFM systems could be applied, from upland watersheds, to logged-over concession lands, coastal mangrove forests, and national parks. Part VI provides a summary of reflections on CFM transitions.

PART II

Continuity and Change in Asia's Forests

Under the increasing pressure of human use and population growth, scientists estimate that Asia's forest cover has shrunk by 70 percent over the past 8,000 years, and that 95 percent of intact closed forest has been lost.[15] The greater part of this deforestation has occurred in the past century. Between 1900 and 1989, Southeast Asia's forest area declined from 250 million hectares to 60 million hectares and is continuing to erode at well over one million hectares per year.[16] In many Southeast Asian nations, forestry policies are being changed in hopes of stemming the further loss of remaining tree cover, but is it too little, too late? According to the findings of the World Commission on Forests and Sustainable Development:

> *Altogether, some 350 million of the world's poorest people depend almost entirely for their subsistence and survival needs on forests. A further one billion poor people—20 percent of the world's population—depend on remnant woodlands, on homestead tree gardens, and on agroforestry systems for their essential fuelwood, food, and fodder needs.*[17]

As these statistics indicate, the health of the planet's remaining forests and the quality of life of some of the world's poorest people are closely intertwined. But, unfortunately, institutionalized approaches to rural and forest development have only in the last few years appreciated that local communities and indigenous people can sustainably manage their local resources. Developmental economists of the 1960's, '70s, and '80s tended to view forests as natural resources to be exploited in order to trigger economic growth in the short-term, rather than as a foundation for long-term rural livelihoods and an essential component for critical environmental services. Foresters clung to their dual role as protectors of the "state forest domain" and professional timber managers.

Despite their orientation towards commercial logging, whether managed by state enterprises in socialist countries or by private companies in capitalist nations, by the late 1980s timber revenues flowed largely into the pockets of small elite groups within the society, while governments were unable to capture rent from timber harvests for investment in social development. While logging failed to generate substantial revenue streams for national governments, it did rapidly drive the deterioration of upland watersheds that resulted in clogging streams, irrigation canals, and reservoirs with silt. Timber operators also introduced large labor forces and roads into very distant areas, opening up the forest to exploitation.

Once logging operations ended, the now jobless, ill-equipped population used fire as its greatest tool to clear land for unsustainable use, accounting for the vast expansion of *Imperata cylindrica* in much of Southeast Asia. Indigenous and upland communities were primarily impacted by the resulting environmental degradation as their source of livelihood and natural resources was eroded. Downstream communities including farmers, industries, and city dwellers, suffered secondary consequences such as increased flooding, lack of clean water, and toxic runoff.

Ironically, the willingness of governments to consider vesting rural people with the authority to care for local forests comes at a time when most of the lowland forests of Southeast

Asia have already been lost. Further, the critical upland watersheds of mainland and insular Southeast Asia are rapidly degrading. That more of these montane forests were not logged in the past is largely a result of their physical isolation, rather than thoughtful conservation policies. Given the rapid and continuing loss of the remaining forests of Southeast Asia, community-based forest management represents more of a strategy to stabilize and restore the region's degraded forests, rather than as a tool to conserve the once rich, bio-diverse tropical rainforests. But, even in the few still pristine forests, the role of communities in management is not just an option, but a necessity.

In many Asian countries, forest departments have primarily engaged communities to restore degraded forests, while much less effort has been given to involving communities in the management of forests with high timber and biodiversity values. This has limited the impact of many government community forestry initiatives in addressing the livelihood issues of poor, forest-dependent communities, while by-passing opportunities to improve management in protected areas and other high value forests. This tentative approach to community forest management on the part of many Asian forest departments represents a lingering mistrust of rural and indigenous populations as well as a reluctance to share rights, responsibilities, and economic benefits in areas with commercial value.

Over the past three centuries, a grassroots movement against the ongoing process of forest nationalization has continued to unfold in Southeast Asia. It is the effort by indigenous people to regain stewardship of their forests that are part of an ancient ancestral domain, of farming communities to protect their homes and forests in endangered watersheds, and of migrant settlers to stabilize their new environment. It is a struggle against logging concessions, state forestry corporations, large plantation holders, mining companies, dam construction, and lowland migrants. And, perhaps most importantly, it is a striving within the communities themselves to strengthen their natural resource management practices towards more sustainable use that will produce greater livelihood and land security. For many forest-dependent communities, the loss of their forest is perceived as the end of their own lives, so closely do they depend on its resources. Just as the loss of habitat threatens the existence of plants and animals so, too, does the loss of the forest endanger the lives and livelihoods of millions of forest-dependent people throughout Southeast Asia.

Land clearing in Northeastern Cambodia. Infrastructure development and consequent in-migration from neighboring provinces result in extensive land clearing and loss of timber resources. Policies and programs on land clearance for agriculture as against logging for commercial production are not yet coherently harmonized in much of Southeast Asia.

While hundreds of millions of people suffer directly or indirectly from forest degradation, their voices often go unheard because powerful interests within government and the private sector often exercise immense controls over resource allocation and use, especially on lands designated as "state forests." Whether logging is considered "illegal" within the legislative framework or carries the "legal" label, most of the industrial timber operations in Southeast Asia occur on "contested domain" where the basic rights of local people are ignored. It is not always a quiet struggle; it is at many times highly vocal and even violent.[18]

Across the region, community forest management initiatives do not represent a uniform movement, but a diversity of efforts and a myriad of forms responding to highly varied contexts. Yet, they share common themes including community discussions and organizing, clarifying and strengthening forest use rules, patrolling and protecting forests, and communicating to outsiders, whether neighbors, government officials, or private companies, that the community has imposed its authority over the forest and is committed to protecting it. The prospect of regaining control of local forests is a source of hope for many rural communities. It represents an opportunity for voices long unheard to be formally recognized and acknowledged.

Over the past two decades, a groundswell of support from government, development agencies, and NGOs has emerged assisting communities in establishing management over their forests and woodlands. Most of these communities would receive high priority as beneficiaries if there was a genuine implementation of the Millennium Development Goals.[19] Planners have crafted national Community Forest Management (CFM) policies, while legislatures have passed laws empowering communities and local government with resource stewardship rights and responsibilities. Development agencies have invested hundreds of millions to support CFM implementation, while urging governments to give community forestry high priority. Scientists have documented indigenous systems of resource use, customary laws, and the long history of resource conflicts. NGOs have multiplied, many focusing on building CFM support capacities including community organizing strategies, participatory mapping and planning procedures, livelihood enhancement schemes and the development of support services.

CHAPTER 1

A Brief History of Forest Management

Two thousand years ago, Southeast Asia was densely forested with scattered communities located throughout the region. Early settlers included hunters and gatherers, shifting cultivators, rice farmers, and coastal fishing communities. Anthropological research indicates that for centuries many of the ethnic communities controlled land and other natural resources through various forms of communal tenure. Utilization rights were typically allocated to households for limited periods of time.[20] Communal resources usually included rights to utilize forests, lakes and streams, grasslands, and quarries.

The broad river plains were lightly populated until the 9th and 13th centuries when pre-colonial Southeast Asian rulers put flood control measures in place and developed a capacity to produce rice surpluses. This early agricultural revolution included irrigation structures, terracing, double cropping systems, buffalo-drawn plowing, new rice varieties, and transplanting techniques. In the pre-colonial era, kingdoms emerged primarily in centers of lowland agriculture or in strategic trading centers such as the riverine plains in the upper reaches of the Chao Phraya basin, and the Irrawaddy and Mekong Rivers, as well as in the Red and Black Rivers deltas of northern Vietnam and the Brantas River in East Java.[21]

In Bali, there is evidence of complex irrigators associations (*subak*) operating as early as 1022.[22] The dominant lowland ethnic community often defined the emerging nation, as did the economy which was usually focused on irrigated rice agriculture. As sedentary agricultural systems were developed that required intensive labor in the construction of terraces, irrigation canals, field bunds, and other work, the concept of individual ownership of land and land sales became more common. The uplands were typically the domain of other ethnic communities, usually a minority, who practiced long rotation swidden farming and were engaged in the collection of forest products.

Cambodia has a long and rich tradition of forest management with carbon dating documenting human settlements as early as 4200 BCE. Inhabitants were likely engaged in hunting, fishing, and gathering. By 1000 BCE, Chinese records report Cambodians were living in houses on stilts and were engaged in cultivated rice production. Upland forest dwellers were likely practicing swidden farming. Most of this early experience with resource management has been lost as it was based on indigenous forms of forest use and stewardship. During the Angkorian Period, considered to have lasted from 802 CE until 1431 CE, the society was oriented towards rice cultivation and fish production from the Tonle Sap (Great Lake). Some scholars believe that the Angkorian Empire declined after its elaborate irrigation systems began to silt-up due to deforestation in the heavily populated areas to the north and east of Angkor.

For over two thousand years, a lively export trade existed between Southeast Asia, China, South Asia, the Middle East, and Europe and it was largely comprised of forest products including aromatic woods, resins, nuts, spices, bird's nests, rattan, precious gems, honey, and other goods. The earliest known evidence of Southeast Asian trade is that of cloves from eastern Indonesia discovered in a pantry in a Mesopotamian town as early as 1700 BCE. Until the 15th century, it is clear that most of Southeast Asia's export products were collected in the forest, rather than grown, including cloves and nutmeg. Of the 43 Southeast Asian trade commodities listed in an account from the mid-13th century (Chau Ju-kua), only pepper and cotton appear to have been cultivated.[23]

By 1400, rapidly increasing international market demands and trade opportunities created incentives for cultivating a broader range of forest products. Between 1402 and 1424, the Ming emperor Yongle sent six hundred ships to Southeast Asia, and tens of thousand of men to procure goods for the Chinese market. By the early 1600s, Europe was importing approximately 5000 tons of pepper annually, as well as hundreds of tons of clove and nutmeg. It is estimated that by the 19th century, nearly one-half million hectares of forests had been felled in Sumatra alone for pepper gardens.[24] In the later 19th century, however, rubber, tobacco, tea, and sugar cane placed far more pressure on forest lands in Sumatra, Malaysia, Philippines and Java.

Prior to the expansion of export crops into forest lands, the major human-induced forest clearing in Southeast Asia was initially concentrated in areas with greater seasonal rainfall fluctuations. This was likely due to problems encountered by shifting cultivators in clearing and burning moist tropical rainforests that had little or no dry season. As a consequence, parts of Luzon, the Lesser Sundas, and South Sulawesi, and the Toba Batak area of upland Sumatra experienced large scale deforestation from the 17th to the 19th century. Land clearing was related to the felling of sandalwood forests of Sumba and Timor, efforts to create more pasture land for wild game and horses, and the continued expansion of areas for shifting cultivation.

The science and framework for state management of forest lands largely evolved during the colonial period. Pre-colonial Southeast Asian kingdoms largely focused their administrative functions on lowland rice growing and sedentary farming areas that surrounded the centers of power. Outlying forest lands were typically viewed as wild areas inhabited by primitive peoples with whom the kingdom would attempt to retain an alliance and trading relationship. Generally, the value of the resources in these remote forests did not warrant the expense incurred in military dominance or the administration of them. While Southeast Asian rulers sometimes claimed control over vast areas of forest land, they rarely ruled them directly and, at most, received some revenue or taxes from inhabiting tribes. For millennia, the forests of Southeast Asia remained under the control of highly diverse local communities, each with their own systems of tenure, customary law, and technologies for utilization.

As a consequence, community management represents an important approach in the history of the stewardship of forest resources in Asia. Unlike their neighbors in the lowland plains and deltas, most upland communities managed the resource communally, rather than dividing the forest land and privatizing it. With vast areas of forests extensively utilized, it was more practical to oversee the resource as a group. Many indigenous resource utilization strategies required a collective effort, including certain forms of hunting, fishing, and swidden farming. Since communities found that swidden required long fallow cycles, rotation of agricultural fields, and periodic movement of settlements, private ownership made little sense. The costs of creating individual forest management units are high relative to communal modes of management. According to Bruce, "[i]n forestry, there are protection and management costs and opportunity costs associated with long-term investment in trees that can more easily be borne by a community or other group than by households."[25]

Community forestry represents a change from "open access" situations, where there is an abundance of resources relative to users and no need to regulate use. As populations expand and resource pressures become more intense, communities find that it is necessary to regulate resource use by community member households. In the past, external pressures from neighboring communities or lowland kingdoms might require that land and forest resource access be negotiated. More recently, pressure from outsiders may be in the form of new timber concessions, plantation estate developers, or migrant farmers which can also drive local communities to assert their forest claims and establish forest protection activities.

In the last forty years, Asian governments have been vigorously extending their claims to "public" forests that are historically occupied and utilized by local communities. As competition for resources grow among government, private sector, and community stakeholders, an increasing number of communities find it necessary to articulate their claim, and secure greater tenure security, through community forestry programs or legislation. In some cases, such as northeastern Cambodia, CF agreements are seen as legal instruments to slow the alienation of indigenous land. In such cases, community forestry provides a tool to gain government acknowledgement of indigenous peoples' formal

Forest Resource Use in Colonial Period. Transporting of logs by aerial trains for distances of two miles. Benguet, Mountain Province in Northern Luzon, circa 1930s, from the album of JK Evans.

management rights for communal land, in the face of growing attempts of outsiders to privatize and ancestral forests.

To be successful, community forestry systems must be able to exclude other users, often generating conflicts in the process.[26] As resource competition grows, communities may continue to elaborate their rules and regulations, including codifying them and even seeking government recognition to enhance their legitimacy. While customary forms of resource stewardship are rooted in pre-history, they continue to play an important role in managing the forests, land, and water in many parts of the world. According to an FAO study:

> *Community-based legal systems, including tenure systems, derive their authority from local custom. These community-based tenure systems, which vary from one ethnic group to another according to land use even within ethnic groups, continue to prevail in most rural areas of developing countries. These systems are commonly called, "customary," or "indigenous," or "traditional" land tenure.*[27]

Indigenous laws might be broadly distinguished between those of rural communities which vary by ethnic group, region, and resource context, and those of Asian courts. The authority of court law emanated from the ruler, who was usually located in lowland agricultural centers, often in major river deltas, or in coastal trading towns. Court law governing property generally dealt with agricultural land. Forest lands, while frequently claimed by the court, were often under the administration of indigenous communities who managed it through customary law. Indigenous peoples typically maintained a relatively high level of autonomy from courts, though they were frequently involved in trade and sometimes paid tribute to local rulers.[28] It was only during the colonial period, that indigenous law and national statutory law came into parallel existence as British, Dutch, French, Portuguese, Spanish and later American administrations began extending their jurisdictional authority into the Asia region.

> *Colonial powers introduced Western tenure forms alongside the indigenous systems... There was sometimes explicit recognition by colonial law of customary tenure rights as well, but in other cases customary rights were unrecognized and unenforceable in the colonial courts... Such recognition tended to be limited to farmland, with the state claiming pastures and forests for itself and seeking to extinguish traditional use rights... Across continents and across different national colonial traditions, the clear and consistent theme of colonial natural resource management was concentration of control over those resources in the hands of the state.*[29]

During the colonial period, state infrastructure and bureaucratic organization that was already present in a limited manner was extended. Nationalization of Asia's forests continued as the elite, originally fostered by colonial interests, came to power with the arrival of independence soon after World War II. As part of the nation building process, new government leaders passed new land laws into legislation that further enhanced the control of the state over forest lands, usually at the cost of customary laws. While most governments have only partially implemented new agrarian laws, and while communities have often resisted them, many of the land laws passed between the late 1940s up to the late 1980s have eroded indigenous systems of resource control and management. This erosion of customary forest management systems has generally been associated with a deterioration in forest stewardship in forests in many parts of Asia. In a study for the World Bank, Bromley and Cernea reported:

> *The dissolution of traditional local institutional arrangements has not been followed by the establishment of more effective institutions, and national governments in most developing countries have not adequately substituted for these former resource management regimes.*[30] *While state sponsored agencies, such as forest departments and state forest enterprises have been authorized as resource managers, or have delegated these responsibilities to private sector timber companies, they have generally failed to implement management rules on the ground that lead to sustainable use. The rise of state agencies and private companies as forest managers has generally coincided with an accelerating loss of natural forests throughout the Asia region during the post World War II era. In Southeast Asia, tropical rainforests receded from 250 million hectares in 1900, to under 60 million in 1989.*[31]

By the 1980s, the deforestation of Asia lowland forests, as well as the deteriorating condition of many upland watersheds, began generating concern among national planners and the development community alike. Floods and brownouts affecting Bangkok, Jakarta, Manila and other urban centers brought deforestation issues to the attention of the public as well, initiating a new generation of environmental protection policies including logging bans. Through the 1990s, there has been growing recognition in many Asian nations that rural people have an important role to play in managing and protecting forest lands, including those nominally under state jurisdiction. "Each year more nations are approving initiatives that provide forest user groups with greater rights and responsibilities in the care of protected areas, upland watershed forests, production forests, and timber concessions."[32]

This historic transition reflects an important shift in forest policy characterized by an emphasis on building state authority from the mid-19th century, to a greater stress on devolution, decentralization, and community rights over the past decade. While it is encouraging to witness the placement of laws and policies supporting greater community and local government participation in resource governance, there is also considerable resistance within government.

CHAPTER 2

Farmers in the Forest

Although the forests of Southeast Asia were sparsely inhabited by scattered groups of people for thousands of years, that scenario has changed dramatically in the past century. This study estimates that there may be over 120 million forest-dependent people in Cambodia, Indonesia, the Philippines, Thailand and Vietnam, representing about one-third of the population in those nations. This estimate includes individuals who live on or near forest land and are dependent on it for a significant portion of their livelihood requirements (Table 1). It is not surprising that forest-dependent populations are large considering that "state forest lands" represent over 50 percent of the national land area in Cambodia, Indonesia, and the Philippines, and over 25 percent in Thailand and Vietnam. Further with much of the forests in upland areas, where sedentary agriculture land is limited and off-farm jobs few, communities are necessarily financially dependent on forest lands for their subsistence.

Given the immense populations and their economic dependence, it is remarkable that the region's governments did not consider them to be a major component in management until recently. Instead, in the 1960s and '70s, communities and their forest use practices were condemned and programs to resettle them gained popularity throughout the region. The eradication of swidden farming, a major forest use system throughout upland Southeast Asia, was a high priority. Since the forest lands on which they lived were nationalized and largely considered "state domain," forest-based communities had no tenure status and were frequently considered "encroachers" and subject to legal action and expulsion. While industrial logging was widely believed to be the engine of early economic development in post-war Southeast Asia, deforestation was blamed on illegal occupants of the forest and the practitioners of "slash and burn" agriculture. Throughout the 1970s, the government attitude towards forest-dependent peoples was hostile, and most development agencies readily supported projects to accelerate logging and remove the unwelcome populations.

Long rotation forms of agriculture, often referred to as swidden farming, are often perceived as one broad category and referred to pejoratively as "slash and burn." Yet, swidden farming practices are some of the earliest forms of cultivation. They exist in a myriad of forms suited to different environments, labor availabilities, and markets. Swidden systems are typically used in upland environments, often in sites where marginal soil fertility limits the number of cropping cycles to several years, after which yields fall steadily. The inability to sustain agriculture is usually due to rainfall, slope, and soil conditions that allow nutrients to rapidly leach out of

TABLE 1: Population of Forest-Dependent Peoples

Country	Total Population 2003* (million)	Number of Forest-Dependent People (million) and Percentage** Total Population	Total Forest land (million of hectares)*** and Percentage of Total Land Area classified as "State Forests"	Ratio of Forest-dependent Population to Forestland
Cambodia	13	1.8 13%	10.5 60%	1 : 5.8
Thailand	64	11 17%	30.1 59%	1 : 2.7
Indonesia	215	65 30%	135.9 58%	1 : 2.1
Philippines****	81	17 30%	15.8 53%	1 : 0.66
Vietnam	81	27 34%	10.9 34%	1 : 0.4

* ADB Key Indicators 2004: Poverty in Asia, p. 62.
** Percentage of forest-dependent population drawn from IUCN WG-CIFM Communities and Forest Management in Southeast Asia (1998). p. 36
*** Total Forest land based on: APFC State of Forestry in Asia and Pacific 2003. FAO 2003. p. 66. (forest cover statistics Indonesia and Thailand)
**** State Forest land is now prescribed as 51% as of 2004.

the soil once forest cover is removed. In such situations, local populations frequently adopt cultivation strategies that involve clearing forest areas and burning vegetation to release nutrients, planting crops for one or two seasons, and then fallowing the land to allow forest succession to proceed along with the return of fertility.

Swidden systems vary widely in terms of the types of plants grown, the variety of crops planted before the site is fallowed, and the period of time the land is allowed to rest and regenerate before being reopened for cropping. In some swidden systems, once the cropping cycle is completed the land is abandoned permanently and new forest is sought, however, many swidden practices involve returning to the same plots of regenerating secondary forest after ten to thirty years. This allows communities to have greater permanence and avoids the heavy labor requirements in clearing old growth forests.

Indigenous forest farmers usually plant a variety of species within a swidden field, diversifying risk and providing a mix of subsistence food and, increasingly, cash crops as well. A study in the Philippines showed that over 30 different crops may be grown in one plot.[33] Many traditional shifting cultivation systems in Southeast Asia have been developed to minimize erosion and fertility loss, ensuring healthy and rapid forest succession and allowing for future agricultural use. These cultivation systems are frequently included with the agricultural practices of migrant peoples who move into upland watersheds to clear forests for crop production. Migrant farmers often clear logged-over forests to plant commercial crops, often as a monoculture system. Migrants may lack local knowledge regarding indigenous strategies to reduce soil erosion and control forest and soil damage from burning. They may attempt repeated cropping cycles, severely depleting soil fertility and constraining future forest regeneration on the land.

There is no question that swidden cultivators place pressure on Southeast Asia's natural forests, though in many areas in the past the cycle of forest clearing and natural regeneration has been remarkably sustainable. For example, a longitudinal study of forest cover in Ratanakiri Province in northeastern Cambodia showed little change in total forest cover over a fifty-year period, only a rotation in the mosaic of land use systems.[34] As rural populations have grown and land areas have remained, static forest clearings have expanded and fallow periods have shortened. Traditional systems that

Farming in the forest. A policy framework that systematically addresses the tenure needs of swidden cultivation is far from activated in any Asian nation. In some areas, rotational swidden farming is evolving into more permanent agroforestry system.

may have allowed 10 to 15 years for natural forest regeneration to take place are currently clearing young secondary growth after 5 to 7 years or less. Upland farmers are compensating for the lack of fallow time for soil fertility to be restored, by adding increasing quantities of chemical fertilizers.

Yet, in many areas swidden fallows are down to three or four years, and are recognized to be inadequate to allow for the recovery of soil fertility. This is especially true where government resettlement schemes, including "sedentarization programs" carried out in Laos and Vietnam in the 1970s and 1980s, relocated upland communities in areas where there was insufficient land for households to establish a viable rotation cycle. Where fallows are insufficient for restoration of soil fertility, lands are eventually abandoned and are colonized by *Imperata* grass and scrub bamboo, permanently losing their productivity in the agricultural system. Such situations, often exacerbated by poorly designed government programs and commercialization, are often used by the same governments as evidence of the weaknesses of swidden farming.

In other areas, rotational swidden farming is evolving into more permanent agroforestry systems. Once initial food crops (tubers, corn, rainfed rice, etc.) are grown, perennial cash crops are allowed to dominate and develop into forest gardens that can provide a long-term source of subsistence foods and spices, fruits, and other goods for sale. This shift from more subsistence oriented swidden farming and fallowing to sedentary agro-silvicultural systems is an ongoing process in many upland regions of Asia. For the most part, this transition is being driven by farmers and market forces, with limited support from government agencies or development organizations. This land use transition has many attractive features including stabilizing forest cover, reducing land disturbance and soil erosion, improving hydrological function and biodiversity habitat.

The transition in resource production systems is often constrained by land tenure insecurity, as very few swidden plots have been titled or even recognized by most governments in Southeast Asia. Typically, land laws respected private ownership only on lands that were historically under sedentary agriculture, while rotating swidden fields were viewed as part of the "state" forest domain. In some Southeast Asian nations, recent policy reforms are giving greater acknowledgement to ancestral lands of upland peoples that practice shifting cultivation, but a policy framework that systematically addresses the tenure needs of swidden cultivators is far from articulated in any country in the region.

The unclear tenure status of upland communities has also created opportunities for lowland migrants to flood into forested watersheds in search of agricultural land. They often bring with them more commercially oriented, monoculture farming systems from the lowlands. In many areas with steep slopes and poor soils, these practices have proved to rapidly degrade soil fertility and generate high levels of erosion. This pattern has been well-documented in the Philippines, where 4 to 5 million indigenous peoples have been joined by 12 million migrants over the past fifty years. Migrant peoples often followed the commercial loggers, entering on their dirt roads to open fields on marginal upland slopes. The pressure frequently displaces indigenous peoples, pushing them further up the mountain and into more critical forest areas. This pattern also exists in Vietnam, Cambodia, Thailand, and Indonesia. From that standpoint, community forest management has broad implications for land law and tenure rights and responsibilities throughout Southeast Asia. It is not simply a question of forest management, but it is tightly tied to human rights and social development concerns.

CHAPTER 3

Industrial Logging, Estate Crops, and Forest Conversion

While commercial timber exploitation was a significant economic activity in Southeast Asian countries during the colonial period, it expanded rapidly after World War II. Growing international markets, improved shipping and processing, and more efficient extraction technologies funded by large capital investments drove the region's timber industry. Viewed as the engine of economic growth, national governments provided large concessions and incentives to attract investors, but an overdeveloped capacity of processing plants increased demand and wastage while reducing the actual value. The deciduous forests of Thailand and the lowland evergreen forests of the Philippines were actively exploited for over a century, with Malaysia and Indonesia taking over as the major timber exporting nations in the 1970s and '80s. Cambodia and Laos entered the sector in the 1990s, and Cambodia is already acknowledging the uncontrollable excesses of its timber concessionaires.

Southeast Asian hardwood timber dominated international markets for much of the post-World War II era, due to the high concentration of commercial species, the considerable value of *dipterocarpaceae* for timber, and the unusually high stocking levels in Southeast Asian forests and teak as a fast growing hardwood. While many African and American rainforests possess a timber volume around 30 m^3 per hectare, Borneo forests average 80 to 100 m^3 per hectare, and some forests in the southern Philippines are up to 300 m^3 per hectare.[35] Industrial logging reached a climax in the late 1970s, as the rich forests of Borneo and Sumatra began to be exploited. Despite the high stocking levels, extraction was extremely inefficient with an estimated 50 percent of the forest destroyed to harvest 10 percent of the timber.[36] Across the region, the timber industry was characterized by extreme inequities. In Indonesia, it is estimated that 15 corporations controlled 578 concessions covering 60 million hectares during the 1970s, including 12 million under the personal control of Prajogo Pangestu, president of the Barito Pacific Group.[37]

According to national forest statistics between 1980 and 2000, many Southeast Asian nations lost from 2 to 48 percent of their forest lands (Table 2). In evaluating the rate of forest loss in Southeast Asia, however, figures generated by national forestry agencies and international organizations are often misleading. The appalling process of forest deterioration and fragmentation is often not reflected in the statistics of the total area under forest cover. Forestry agencies are under national and international pressure to show they are slowing deforestation. As a consequence, statistics concerning forest cover are often inflated, and at best indicate areas with some continuing level of forest cover, not the condition of the forest. Another perspective on deforestation in Southeast Asia is presented in the World Resources Institute report on "frontier forests," which are defined as relatively undisturbed forests with sufficient size to sustain their original species composition. From this perspective, Southeast Asia's forests are badly depleted

TABLE 2: Forest Cover and Percentage of Frontier Forest Remaining (1980-2000)

	Cambodia	Indonesia	Philippines	Thailand	Vietnam
Forest Cover in 1980 (million hectares)*	13.5	124.5	11.2	18.1	10.7
Forest Cover In 2000 (million hectares)**	9.3	105	5.8	14.8	10.9
Decrease 1980-2000 (%)	31	16	48	18	2
% of Original Frontier Forest***	10	28	<3	5	2

* World Resources Institute: 1998-1999
** APFC State of Forestry in Asia and Pacific 2003. FAO 2003. p. 66 for Cambodia, Indonesia, Philippines, and Thailand. Figure for Vietnam from MARD Statistics 2000.
*** World Resources Institute, The Last Frontier Forests: Ecosystems and Economies on the Edge (Washington, DC: WRI, 1997).

and severely threatened. Of the countries reviewed, Table 2 indicates that only Indonesia possesses more than 10 percent of its original frontier forests. Virtually all of the original frontier forests of mainland Southeast Asia, with the exception of Burma, are gone.

In most Southeast Asian nations, the population has increased approximately 300 percent over the past fifty years, and this has certainly added to rural pressures on remaining forest lands. Yet, it was the opening of remote areas by logging companies after World War II that accelerated their depletion. Forest cover in the Philippines dropped from 80 percent of the land area in 1900, to less than 19 percent a century later. Old growth forests now represent just 3 percent of the land area, though they are too small and fragmented to be considered frontier forests. Current rates of deforestation are dropping primarily because of the limited area left and its inaccessibility and lesser volume per area. What remains is montane or mossy forest, as all lowland forest has been exploited. A World Bank estimate indicates that at the current rates loggers will deplete virtually all of the lowland forests of Kalimantan by 2010.[38]

Given the damage done to the natural forests of Southeast Asia during the 20th century, the era of commercial timber extraction is drawing to a close due to the lack of standing stock and the certification of sustainable extraction that is being demanded in global trade. Much of the timber concession areas are now being converted into commercial estate crops. Oil palm, coffee, rubber, and other plantation crops already cover millions of hectares and are expanding at a rapid rate as national and international investors seek new opportunities for profits. The formation of a joint Ministry of Forests and Estate Crops in Indonesia in the 1990s both recognizes and facilitates the ongoing conversion of state forest lands to corporate farms, especially on the islands of Kalimantan and Sumatra. In the Central Highlands of Vietnam, the rapid expansion of coffee plantations in forest areas reflects a similar trend. There is still little effort to enhance secondary forest and the rush is evidently on to log by what means possible. Where logging is seen as part of the needed forest land conversion to spur national agricultural productivity, governments often vest this with legality even though its legitimacy at the local level is compromised by lack of stakeholder participation. [39]

Forest fires have strong ties to plantation establishment and are a major factor in accelerating forest loss in Southeast Asia, especially over the past decade in Indonesia. In 1994 alone, some 5 million hectares burned in Indonesia.[40] Satellite imagery from 1997-1998 indicate that during that fire season the majority of burns were caused by large-scale land clearing for pulp and oil palm plantations. An FAO report indicates the total area burned in Indonesia in 1997-1998 at 9.7 million hectares with the economic loss was between $4.5 and $10 billion and 75 million people affected by fire, smoke or haze.[41]

These fires are of concern not only for their extent, but their intensity. Fires in the rainforests of Kalimantan are an unusual occurrence and those of the past decade have been particularly devastating as they were often fueled by slash left from past logging operations and coal seams near the surface that were ignited and continued to burn under the ground through the rainy season. Recent studies indicate that many of these fires will likely result in the permanent conversion of natural forests to non-forested land. In some cases, this results from the extreme heat of fires burning in logged over areas with high fuel loads that destroy virtually all canopy trees, sterilize soils, and create environments where natural forest regeneration will be constrained.[42] The steady degradation of forests in Southeast Asia through logging, fire, and land clearing continues up to the present, despite the gradual deceleration of industrial logging.

Illegal forest activities have significantly contributed to forest depletion. Illegal activities are further upheld by poorly implemented policies for sustainable forest management, the inadequacy of monitoring and evaluation, and the lack of initiatives to rectify forest land allocation, thus illustrating some of the limits to centralized management of natural resources. Motivation to address forest management flaws mainly comes from social outcry rather than from internal reviews of concessions.

CHAPTER 4

Social Responses to Deforestation

The appalling loss of Southeast Asia's forest over the past fifty years has not gone unnoticed. Villagers are well aware of the impact of their own and neighbor's use on forest cover, as well as the effects of logging operations, plantation establishment, dam construction, resettlement projects, and mining activities. In communities living within and around forest areas, deforestation has often had a profound impact on quality of life and income generation. Economic pressures have often forced communities to participate in illegal logging, while others engage in such practices in the hope of improving household livelihood, despite a growing awareness of the long-term environmental costs.

Urban residents are also not untouched by forest loss. As downstream flooding and power shortages grow more frequent and severe, middle class city dwellers have begun mobilizing political pressure to force governments to take action, although seldom in support of upland communities. This expanding environmental movement has its strongest support among urban middle-class people, who are increasingly well informed through the mass media. The mass media, however, tends to highlight problems and report on environmental disasters creating great concern among its audience without necessarily informing the public about the possible solutions at the policy or field level. Further, the perspective of the rural poor vis-à-vis the environment is often poorly presented through the media and in national policy debates in most Asian nations.

Media has effectively convinced the public that logging is an irresponsible industry and upland migrants also contributor to watershed degradation. Some sectors view logging bans as expedient in halting deforestation and conserving forests. Other sectors propose sustainable logging of natural forests to meet the national demand for wood that plantations cannot yet provide.[43] The public generally views deforestation as the main cause of flooding disasters. Basically, the complexity of the social problem and the accountability in different sectors of society are yet to be mainstreamed.

Commercial operations of the past decades left behind logged-out areas and jobless migrant employees who stayed on to till these abandoned and open access areas for lack of alternative livelihood. Skilled in logging and land clearing, these people became ideal contacts for private interests seeking cheap logs without responsibility for replanting. If local responsibility was immediately harnessed for managing logged out areas, then forests could have regenerated after commercial logging operations ceased.

The question then is not whether a logging operation could sustainably manage an area but rather, can it be viable under present circumstances and can it be reproduced in different areas? The Philippine Wood Producers Association, for example, acknowledges its tough financial and social predicaments. The industry needs finances to prove that it can measure up to criteria and indicators for sustainable management and dispel the public perception that there is no way that logging can be sustainably done. This is made more difficult as clarity of operations is not easily achieved nor transparency of transactions granted given the diversity of licensing agreements having different terms of management signed under different administrations.

Logging bans are often issued to show that government responds to disasters blamed on deforestation. Yet, effectiveness of its implementation as a national policy has been in question.[44] There are too many loopholes by which destruction of the forests can continue, thereby reducing the credibility of logging bans to political and emotional reactions during a state of hopelessness. Log ban policies, where they do exist, have proven to be inadequate and inconsistent for the management of existing forests. For countries that are unable to protect themselves from surrounding destructive activities, people express doubts as to whether their governments can effectively enact policy decisions such as logging bans. They view such decisions to be biased by political pressures from neighboring countries or by the immediate economic gains to be made locally. Though ITTO and other international institutions may justifiably argue that sustainable logging is possible, many citizens are yet to be convinced that their governments are capable of its implementation. While logging bans may not be successful today in protecting forests, the repeated push for this policy is saying something about society's lack of confidence in government's ability to fulfill its role as much as the logging industry itself.

Tree plantation development is presented as another strategy to reduce pressure on natural forests and meet wood demands. Yet in some cases contractors do so through initially clearing secondary forests, thereby destroying what can be regenerated. While domestic demand for timber has often been expressed as a concern, production has not been geared towards meeting the volume and type of wood needed within many Asian countries. The domestic need is usually for lumber but foreign investment coming in is primarily for pulp that can be exported. Wood that ends up on the local market faces too many restrictions and disincentives, along with low pricing and limited processing – hardly justifying such domestic investment in the first place.

While deforestation does reduce infiltration and may significantly affect the seasonal water balance, during major tropical rainfall events not even primary forest can withstand the soil saturation and slippage that cause massive landslides and temporary dams along riverbeds. The resulting floods only turn into disasters when people have occupied the downstream flood channels. These people are usually the urban poor who are left with only unstable and flood-prone areas on which to build their houses. If disaster prevention is incorporated in land use planning and implementation, then safe land should be made available for all housing.

The myriad of concerns over deforestation and the need to find national solutions to growing environmental problems has fostered the establishment of formal civil society dialogues in a number of Southeast Asian countries that draw on the sentiments of rural and urban peoples. These attempts to bring stakeholders together at a national level are taking the form of working groups, special committees, environmental networks, and forums that meet regularly to debate and advise government policy makers. Finally, national governments in a search for more sustainable systems of natural resource stewardship have been strengthening community and local government roles through democratization and decentralization policies and laws that have been promulgated in recent years.

Over the past decade, many Southeast Asian nations have passed laws to devolve and decentralize government administrative functions to the district and sub-district level officials. In principle, local government officials are meant to be downwardly accountable to constituent populations. As such, democratic decentralization is an effort to enhance popular participation in resource stewardship and formally institutionalize it within a governance framework.[45] Unfortunately, this process has had mixed results and local government officials may be more responsive to their superiors than to their constituents, especially if they are appointees. Corruption has not been removed through the devolution of governance systems, but in some cases it has simply been decentralized.

Decentralization often involves new systems for the election of village representatives, delegation of small development budgets, and new authority for the management of natural resources. As part of the process of decentralization, forest department field staff may be increasingly placed under the supervision of provincial or district authorities, with less control by central technical ministries, as has occurred in the Philippines. Arguments in favor of decentralization include greater equity in resource access and greater management and economic efficiency in terms of reduced transaction costs, better matching of services to needs, mobilization of local knowledge, and increased public sector accountability.[46] However, in most countries decentralization is only happening in a limited way, for example, few indigenous people are engaged at the lowest level of governance in most parts of Asia. While local government is given the responsibility to monitor resources, rarely is it empowered with the authority to allocate resources, while having limited revenue and skills to assist communities.

Yet, in the long term, decentralization trends may hold a promise of new support for community-based resource management, especially if they involve democratically elected representatives from the rural communities. With forest management under local government authority, some resource management specialists feel that communities will be better able to negotiate collaborative agreements. For the most part, the impact of decentralization policies on community involvement in forest management is hard to assess, as these reforms are still being implemented. Capacity at the community level, as well as local government level, to formulate and implement management plans is still in an early phase of development. Accountability of local government representatives to constituent communities is often limited, as many provincial and district officers still see their authority as emanating from above, rather than from the communities below.

In other cases, local government chiefs operate as authorities unto themselves, and the decentralization of greater control to the province from the central government has only accelerated deforestation. This has been the case in many parts of Indonesia since the end of the New Order government brought decentralization policy reforms, as well as in some provinces in the Philippines. In India, the social forestry programs of the 1970s and 1980s relied on village *panchayat* institutions to administer forest management activities including woodlots and roadside plantations, with disappointing results. In retrospect, *panchayat* organizations, covering 5 to 20 member villages, were found to be too removed from and inexperienced in forest stewardship to be effective managers. Only twenty years later did program strategy shift to hamlet-centered community forestry.

A recent World Resources Institute study in Southeast Asia noted that community natural resource management strategies may not be supported by national policies, even when NRM responsibilities have been delegated to local government, ultimately undermining local management initiatives.[47] Asian forest departments, that have enjoyed almost unilateral control over state forest lands for decades, are often very reluctant to share their authority with local government representatives. In India, it is possible to interpret the Ministry of Environment and Forest's establishment of the Forest Development Agency (FDA) at the district level, rather than to coordinate with local government planning agencies established under the *Pachayati Raj* Act, as an attempt to create a parallel channel for development funds that state forest departments could control unilaterally.

There is no question that in many countries, different ministries and agencies are struggling for control over forests and forest development budgets. A major contest in coming decades will be the debate between the Ministries of the Interior to grant greater rights to local government and the efforts of forestry agencies to retain their historic rights as "state landlord." In countries where the political elites have utilized partnerships with forest departments to manipulate national forest resources to enhance their power, they may find it easier to work with the technical agencies, than to broker new arrangements with local government. As democracy systems gain momentum in Asia, elected politicians are also using state forest resources as a private source of campaign financing for election to higher office. In some Indian states, Forest Ministers channel funds into party coffers, both through the sale of forest department positions, as well as levies from forest revenues. To that end, some national political powers may prefer to support continued control of state forest land under forest departments, rather than press for decentralization.

In the near term, most fledgling local government entities in Asia are most concerned with infrastructural issues, including public health, education, market access, and roads. Much of the discretionary funds allocated to administrative village, sub-district, and district governments are for infrastructural development activities, and as a consequence these will take priority over marginally funded natural resource management needs. Over time, as local government planning and project capacity expands, they will likely increasingly engage communities as partners in local forest management. There are, however, already a growing number of contexts where local governments are beginning to pay attention to environmental management and resource planning. These tend to be in rural areas with a high degree of forest dependence and forest resources availability, where natural resources are threatened, or where NRM project support is available and being promoted.

While the rapid deforestation of Southeast Asia during the last three decades has recently begun to find reflection in the emergence of community forest management policies, as well as a broader trend towards the decentralization of natural resource management authority, it has also found a voice among the region's rural communities. Villagers in growing numbers are concerned over the alienation of ancestral lands, the degradation of natural forests, and the loss of important sources of livelihood and residential place, and they are organizing and expressing their feelings to local and national government representatives.

According to Jeffrey A. Sayer, past Director General of the Centre for International Forestry Research (CIFOR): "In many countries, the most visible symptom of bad governance has been abuse of forests and land, and the lightning rod for expression of public dissatisfaction with corrupt governments has been the struggle for equity in access to natural resources. It was, therefore, no surprise to find Indonesia's nascent environmental organizations at the forefront of the barricades during those tumultuous days in May 1998 that led to the overthrow of the Soeharto regime."[48] By 1999, the process of forestry sector reform was well under way, and decentralization and community forest management were high on the political agenda in Indonesia. Such reforms have also been high priorities in the Philippines and Thailand at the national political level, while moving forward with lower profile in Cambodia and Vietnam.

The growing popularity of decentralization laws and policies in Southeast Asia, however, does not guarantee the devolution of state forest lands to communities or even a meaningful empowerment of local governments. Forestry agencies have repeatedly demonstrated a reluctance to see their authority shared with other partners. Much of the management responsibility being shifted to community groups and local government involves protection duties, rather than significant access to resources. Where resource sharing agreements are being approved, they often leave the forestry agency with considerable control over harvesting and marketing decision making. As the following experiences indicate, community forestry is part of a long-term transition in forest management that is still in a nascent phase of development.

In the past ten years, many nations in the region have begun crafting new policies and laws that enable communities to gain greater legal rights and responsibilities as stewards of "state" forest lands. This trend is tied to an apparent recognition by some policy makers, development planners, resource managers, and economists that commercial timber exploitation models are proving to be unsustainable, and that the long-term costs will far outweigh short-term benefits. National policies are an attempt by those in positions of leadership to articulate the directions that society is meant to take. They are "attempts" because policies may or may not be implemented, and may or may not influence resource use behavior, depending on their effectiveness. What one hand of government may create the other may destroy; community management may exist as law, policy, and programs, but unless communities are actively supported they may not be able to develop sustainable production systems and marketing capability.

For that reason, addressing social and environmental problems may begin through the passing of legislation or development of a policy, but ground level impact is the only way to assess its efficacy. Changing forest policies also reflects a growing awareness among policy makers that forest-dependent populations are large and expanding, and the short-term economic benefits derived from industrial use do not justify the loss of livelihood experienced by millions as the forests are degraded. As democratic institutions continue to develop across the region, forest-dependent people are generally gaining greater political influence in shaping forest management policies. Part III will explore recent forest sector policy and legislative changes and discuss unique shapes they are taking in each Southeast Asian country.

Part III examines the types of laws, policies and programs that are emerging in Southeast Asia that are permitting the formal integration of community forestry within national forestry sectors. This discussion refers to a *policy* as a course of action, guiding principle, or procedure considered expedient, prudent, or advantageous by the government. The term "law," as used in this report, describes a body of rules enacted by legislation. Policies are usually not ratified or enacted by the legislature, but by the executive branch (the government). Sometimes policies can be within a ministry, providing implementation guidelines for new laws. This section discusses how both policies and laws are providing formal frameworks for community management that allow informal traditional and emerging systems of resource stewardship legal recognition. Part III also explores the extent to which national CF programs, and the funding that supports them, are driving law and policy implementation on the ground.

PART III

Laws, Policies, and National Programs

In some countries, like Vietnam, national policies are a primary factor guiding transitions towards community forestry and decentralized NRM, while the legal frameworks to lawfully endorse such management arrangements are still under development. In other countries, such as Cambodia, laws are being promulgated while policies are being put in place to see that new legislation is implemented. In still other nations, including Thailand and Laos, externally funded projects are the primary force linking forestry agencies and communities in newly forged management partnerships. Ideally, community forestry policies, programs, and supportive legislation should go hand-in-hand to facilitate systematic implementation and should be consistent with related laws and development policies. Differences between the legislative and the executive offices, however, may also generate conflicting policy directions regarding how public forest is to be used. Powerful figures in the legislature may desire to channel forest rent to fill election coffers, while the executive and military may see it more as a source of personal financial gain given the poor compensation gained in public service.

Past forest nationalization policies and laws have typically empowered government technical agencies, state corporations, and private concessionaires to manage the public forest domain. In response to rapid forest depletion throughout Southeast Asia, there has been a recent trend towards policies and legal reforms that engage communities as new partners in state land management. Such policy and legal reforms are often encouraged by external actors who hope they will contribute to a stabilization of the natural resource base, while addressing livelihood concerns.

Since most Asian nations nationalized much of their forest land in the 18th and 19th centuries, placing the management of "state domain" under technical agencies, legal changes are required to authorize the transfer of various rights and responsibilities to communities and local authorities. Countries across the region have handled this issue in various ways. These include promulgation of new laws and regulations, policies set forth by technical agencies and the political leadership, as well as judicial decisions taken by high courts. International laws, treaties, and agreements, as well as policies of international development agencies can also influence national laws, policies, and strategies that address community forestry issues. The following chapter reviews some of the legal actions taken and processes that are being used to formalize and develop national CF strategies in Asia.

CHAPTER 5

Laws and Legislative Process

There is a general hierarchy of policy and laws that provide a formal governmental framework for society (Table 3). At the highest level, social policy is framed by the constitution which is typically formulated by a special committee. Laws are drafted and ratified through the legislative branch. The office of the executive branch, including the prime minister or president, a ministry, or the governor of a province or state, promulgates regulations. A district or sub-district government or local village administration may formulate local rules. Finally, the judicial branch of government makes court decisions. Any one of these processes can lead to actions that influence the authority of communities over certain state forest resources. Throughout Asia, these institutions are taking actions that influence the role of communities in managing forest lands. Typically, high-level policies and overarching laws are framed and articulated in a general manner, allowing for the development of more specific regulations and rules at the operational and local government level. If these operational orders and decrees are viewed to be inconsistent with the overriding laws, they may be challenged in court, allowing the judiciary to determine their consistency with the intent of broader documents such as a country's constitution or various natural resource and land use laws. New laws need to be drafted consistent with the existing body of national legislation, unless the law being drafted is to replace existing law. Consequently, nations must work within their existing legal framework to craft laws addressing community forestry issues. In Southeast Asia, many nations continue to be influenced by their experiences during the colonial period that affect their legal environment. In the Philippines, Spanish and U.S. legal influences persist, while Indonesia has a Dutch legacy, and Vietnam and Cambodia a French one. Aside from their different historical roots and corresponding legal cultures, there are no universal standards that can be applied in order to formulate ideal community forestry policies and laws. An FAO study of Common Property Resources (CPR) policies found that legal approaches to formulating community-based natural resources management necessarily varied widely, due to the diverse contexts for which they were designed.

In other words, policy and legal frameworks need to be designed to fit specific historical, socio-political, cultural, and legal realities. Such policies and laws need to take into consideration both the past history of forest and land tenure laws as well as existing policies. At the same time, new national policies and laws must consider the impact of natural resource behavior due to socio-economic changes within a society. Policies and laws can then help encourage desirable behavior and constrain unwanted activity. While there is no "blueprint" for community forestry, effective laws and policies typically share a number of broad characteristics designed to facilitate community mobilization and empowerment.

Bruce defines three broad sets of rules that, in theory, are fundamental to common property resource management.[49] Firstly, rules establishing collective choice are generally handled by the law of associations. These may include cooperative law, corporate law, registration, and forestry law. These laws provide for: 1) the constitution and legal personality of the community, 2) the delimitation of its members, 3) the authority to control members' activities, and 4) the process of community decision making. Secondly, rules governing management authority are usually broadly framed within constitutional law, finding greater definition under property law, though conservation law, forestry law, and contract law may also determine allowable and

TABLE 3: Mechanisms for Establishing Social Policy, Legal Frameworks, Executive Orders, and Judicial Decisions

Type of Mechanism	Ratifying Institution
Constitution	Special Committee
Treaties	Ministry of Foreign Affairs, Ratified by Legislative Branch
Laws	Legislative Branch
Regulations	Prime Minister/President, Ministry, Province/State
Local Rules	Local Administration or authority
Court Decisions	Judicial Branch

unallowable uses. As new CF legislation is approved, much more specific guidelines are being developed by some Asian countries providing regulations of use by members and non-members.

And, thirdly, rules defining external arrangements, including the relationship between community managers and external groups are another important group. These may include non-members within the community, neighboring communities, private sector interests, etc. Relations with government and community institutions are affected by legislation establishing and empowering government ministries or agencies and by the laws that establish the hierarchy of controls between different levels of government. Among other things, these rules provide for the resolution of disputes between communities and external actors.

A fundamental requirement for formally empowering community forestry is the need to transfer "exclusive" rights and incentives to a specific and limited "membership." As Bruce and Fortman note, it is important to identify and empower "a group with a limited membership, with a right to exclusive use of the resource, which then has the opportunity to regulate resources used by its members and also the incentive to do so, because the costs and benefits of disciplined, sustainable use are internalized to the group."[50]

In the Asia region, community forestry has often been most successful when supportive policies and laws facilitate spontaneous, community-based initiatives to protect threatened forests. A review of community forestry cases by the Workshop on Political Theory has suggested that groups emerge to manage common property when the user population lives close to the resource, is relatively small, faces moderate resource scarcities compared to demand, and where the resource is subject to multiple use requiring management and coordination.[51] The study also found that community groups sustain community forestry management operations where the group has clear-cut rules enforced both by users and officials, internally adaptive institutional arrangements, the ability to nest into external organizations to deal with the outside environment, and different decision rules for different purposes. Policies and laws that are designed to support such conditions are likely to be more effective in enabling community forestry to be established and sustained. As a consequence, many community forestry related policies and laws appear to include common elements that support small community groups by extending clear rights and responsibilities for forest management. Some key elements are included in the following section.

Key Elements in CF Laws and Policies

Most CFM policies and laws emerging in Asia over the past decade have similar content. They include rights of association, tenure, access, to manage; monitoring and enforcement responsibilities; and conflict resolution mechanisms. A brief discussion of each of these elements is provided below:

Rights of Association reflects the legal mechanism through which the institutional entity responsible for discharging the management functions is legally identified and authorized. In the Asian context, community-based forest management is usually carried out by the households in small villages or hamlets that are located in the vicinity of the forest land in question. In most countries in the region, settlement groups below the administrative village level usually have no legal status, presenting problems when formal management agreements are attempted between the government and the group of households in question. As a consequence, the "rights of association" become a major issue in many CFM policies and laws. In some countries, rights of association are conveyed to the group of households when they are registered as a society, non-profit organization, trust, or another form of civil society institution through a security and exchange commission. In other countries, they are formally registered as member bodies under a technical agency, such as a forest department, labor department, or cooperative development authority.
Further, in some cases, the hamlet is empowered through the administrative village, commune, or sub-district that oversees its operations. This right of association may also include recognition of the cultural identity of the people involved.

Tenure Rights articulates the status of the management group's claims to the forest resources. This may include the time period for which the managers possess rights over the forest land and resources, the extent to which extensions are available and under what conditions, whether the land may be inherited by group members, and when and why "rights of tenure" may be dissolved. Since in most CFM policies and laws in the region, primary ownership of forest land is retained by the state in perpetuity, the tenure rights held by communities do not amount to full ownership, but are more likely to take the form of leases or other time-bound contractual arrangements. Such tenure rights do not include rights over minerals except in a few cases where extraction traditionally occurs.

Rights to Access and Manage usually conveys the terms under which products may be collected and harvested from the forest. They may also include rights for grazing, water use, gardening, and recreation. These may include arrangements for sharing benefits with the local government and technical agency. Rights to access include spatial and/or seasonal restrictions. *Rights to Access and Manage* may delineate authority over forest management planning activities including species to be planted and harvested. Requirements for mapping, felling cycles, replanting, soil and water conservation, and transportation and marketing may be included. Such rights may recognize cultural rituals, but not always cultural practices, most especially swidden. Rights generally do not include what is under the land, nor the selling of water resources off the land.

Monitoring and Enforcement Responsibilities are important components of many CFM policies and laws. Typically, CFM strategies are cast as management partnerships between local communities and government bodies. An important role of the communities is forest protection and, as a result, forest monitoring and forest law enforcement are responsibilities delegated to local managers under CFM policies and laws. Under collaborative management systems, forest departments and local government may also share responsibilities for monitoring and enforcement. Local rules and regulatory practices may be recognized as part of the area enforcement and responsibility.

Conflict Resolution Mechanisms are an important component of most recent CFM policies and laws. Since CF strategies are designed to vest specific communities with forest use rights and management responsibilities, it is important that a formal process is developed that allows conflicts over resources to be resolved. Conflicts may arise within the community, with other communities, with the forest department, local government or other stakeholders. Multiple tenure instruments may exist in an area and the processes of resolution may be specifically defined (e.g., Cambodia) or lack a clear and definitive process (e.g., Philippines).

A key element in any CF or CPR policy or law is transferring exclusive rights to the resource. Experience in Southeast Asia indicates that even where policies and laws clearly empower community managers, their rights will frequently be challenged, especially in the initial years after the formal transfer of responsibilities. The extent to which the law enables communities to enforce their forest management rights, and the actions it allows, may influence the communities' success in protecting their resources. According to Bruce's assessment:

> *The creation of a CPR, like the creation of individual property, confers the right to exclude from the resource. That right to exclude is the essence of property regimes... The community's right to exclude will understandably be challenged by others who want access to the resource. Some challenges will come from previous users. Others will come from new claimants, backed by national powers. These interests often target extensively or seasonally used common pool resources, asserting that they are vacant and underutilized... We should expect the legal basis of common property to be challenged from both inside and outside the community. Those challenges most often are not formal legal challenges, but acts that violate rules of exclusion and rules governing use, acts that test the will and ability of the community to enforce its rules and seek to undermine the CPR.*[52]

Another important consideration in evaluating policy and legal mechanisms designed to support community forestry systems is the extent to which they interface well with indigenous or informal community resource management rules and practices. While some CFM systems are based on new community institutions, many draw on past traditions of resource management and indigenous ways of organizing and governing group behavior. As Bruce notes, national statute law developed during the colonial period and until recently has often "undermined the formal position of the community-based tenure systems and interfered with enforcement of their rules, without providing a viable alternative. The earlier systems continue to exist, sometimes modified by attempts to replace them. The result is a pervasive dichotomy between the formal legal situation under national statute law, and the law-in-action, that is the local, community-based system, upon which local people still rely for access to land."[53]

Clan, lineage, and village are traditional forms of social organization that have historically played an active role in the management of forest lands. Traditional territory of such groups, often referred to as *ancestral domain*, is often given little recognition under national statute law, in part because most Asian nations provide little or no legal recognition to organizations below the level of administrative village. As a consequence, membership in a clan, lineage, or hamlet is unofficial, as is the activities and governance activities of such bodies.

Indigenous systems of tenure are often highly differentiated depending on land use type and place. Bruce notes that "To understand the landscape of indigenous communities in tenure terms, it is useful to have reference to the concept of a 'tenure niche,' a discrete area of land within that landscape defined by applications to it of a specialized set of tenure rules."[54] The concept of a *tenure niche* reflects the diverse and flexible ways in which communities form and adapt rules and regulations governing resource use and benefit distribution to respond to local conditions and opportunities. National policies and laws that rigidly prescribe sharing arrangements and generalize across varied production systems and natural and human defined landscapes constrain adaptation, frustrating community efforts to develop operational management systems on the ground. Given the many problems involved in integrating indigenous property concepts and land management institutions into national land law, some countries are exploring recognizing ancestral domain areas and allowing them to be managed under existing customary systems within their designated area. Perhaps the greatest diversity in government responses has been in the Philippines with the granting of ancestral titles where the communities document their traditional practices as part of the accepted management plan.

In other countries in Asia, attempts have been made to formalize community institutions by relating them to existing laws and institutional mechanisms. These laws, however, are often outdated and inappropriate for rural community groups, and as a consequence the models proposed for village-based community forest management committees may complicate, rather than enable community stewardship by introducing legal problems that constrain community action. For example, in India the Societies (Registration) Act of 1860 has been used to formalize CF groups, requiring the adoption of by-laws and institutional structures that do not reflect traditional modes of village organizing. The Act was originally promulgated to provide legal status for urban civic clubs such as garden societies and kennel clubs during the colonial period, and is now used to legally register CF groups and thereby empower rural communities with a legal identity needed to open bank accounts.[55]

One strategy that has been increasingly adopted in the Asia region to formulate better CF policies and laws has been a process of drafting legislation followed by civil society dialogue and revision. This has often helped to inform new policies and laws and make them more responsive to field level realities and needs. Yet, while some Asian governments are exploring ways to gain civil society input into law and policy formulation, disempowered rural communities have limited voice in such dialogues, while powerful government actors and private sector stakeholders remain dominant. Before reviewing country experiences with CF policy development, it is useful to reflect on the generic process of participatory policy formulation.

Process of Law Formulation and Approval

The laws developed in a number of Asian countries to guide community forestry programs often follow a similar pattern of formulation, approval, and revision. A process for drafting is identified, a drafting team or working group is formed, and the drafting process and team are approved by government. Materials are reviewed by the drafting team, including related laws and the social context that has often precipitated the initiative, with consultations and inputs during the drafting process. In some cases, stakeholder input is sought through public hearings and workshops to identify important issues, needs, and concerns prior to

FIGURE 1: The Cycle of Community Forestry Legislation Development

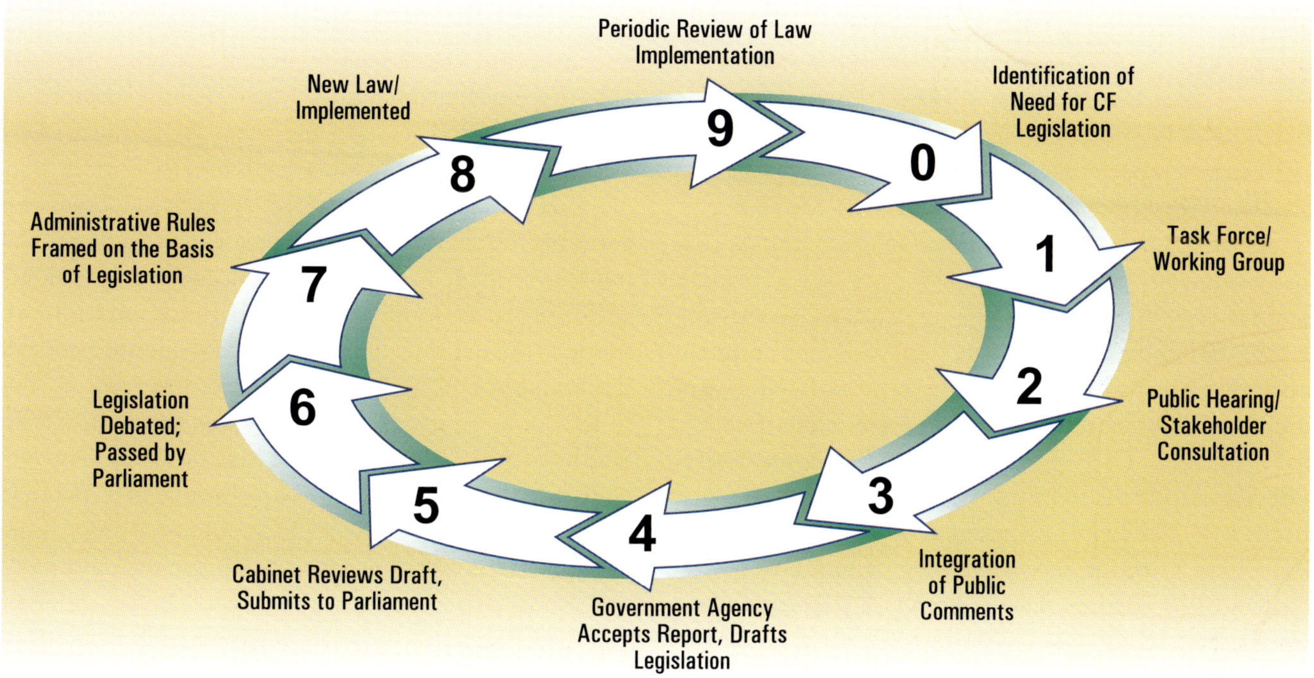

drafting the new law. In the case of Thailand and Cambodia, for example, community forestry legal initiatives have experienced this type of civil society review.

Typically, the national forestry agency takes the lead in drafting the supportive legislation. Once a draft is developed, it may be circulated for public review. After public input has been made and revisions done, the draft law is submitted to the legislative branch for review and ratification. Once approved by the parliament, house and/or senate, the legislation (whether it be a law, decree, sub-decree, bill or other mechanism) returns to the implementing agency for the development of operational guidelines. In the case of community forestry laws, operational regulations usually take the form of implementation orders which are prepared by a technical agency, such as a forest department, for approval by the relevant Ministry. Once the operational guidelines are issued, the law may be implemented. Figure 1 presents a basic description of the steps involved in the process of law formulation and approval. Rarely is it as simple as Figure 1 suggests. Implementation of new legislation is facilitated by policies and programs that provide political support and financing. Periodic reviews may go beyond and call for new legislation keeping the clock moving. In Asia, the effectiveness of community forestry laws are often scrutinized carefully in the initial years and revised as field experience is fed back to policy makers, technical analysts, and legislators. In the Philippines, a variety of legislation has been developed and approved over the past decade to support community-based forest management, while in India Joint Forest Management (JFM) implementation guidelines from the center and the state level have been revised a number of times. Programs are often supported by bilateral grants and loans that have as part of the agreement a condition to develop new laws. Such legislation, however, may not reflect broader national strategy, and may last only as long as the project support.

Limitations of Existing Laws and Policies

There are no universal standards that can be applied in order to formulate ideal CFM policies. After carrying out a similar review a few years ago, regarding CPR policies, Bruce notes:

> *One cannot derive legal forms that are universally "right" for common property forestry management. There are some lessons to be learned, of course, but there is no axiom more basic to the study of law and society than that a legal rule (a command to act or refrain from acting in a certain way) will*

produce different behavior on the part of individuals differently situated. "Differently situated" can mean situated in different economic classes or social groups, but it can also mean situated in different cultural and political milieus. The clear implication is that different legal solutions will be required for different contexts.[56]

Communities require a firm legal basis for managing community property that interfaces well with indigenous and local institutions and prevailing resource use practices. The legal vulnerability of community forest management is one of the fundamental challenges to the sustainability of the resource stewardship systems. According to Bruce, laws governing common property "is treated by statutory law in almost all countries in an unfocused and fragmented fashion. There is no single statute or even field of law that covers all of common property,"[57] and as a consequence "it is often difficult to provide legal security of expectations, including security of tenure, to communities managing land resources under indigenous common property regimes or seeking to construct such regimes."[58]

Bruce identifies four reasons why a firm legal basis for CF is difficult to address: [59]

- Lack of solid legal provisions for CF in national law relating both to organization form and property rights;
- The exclusions of claims to such rights for local communities by the state's claim to own vast areas of the rural land resource, including forests;
- The poor integration of parallel systems of customary and statute law, which may vary between national law, state or provincial decisions, and local communities; and
- Overlapping and poorly reconciled, if not conflicting, systems of law operating concurrently.

National legal statutes build upon prior legal actions, both legislated legal codes as well as court decisions that create precedents. In creating legal space for community forest management systems to operate, new laws and regulations must find a place within the existing legal framework. The specialized statutes regulating natural resource use, typically formulated by forest departments and other government agencies may provide for simplified forms of organization and authority for community management of the resource, but government orders or regulations alone are rarely sufficient to respond to the legal needs of common property management or to deal with tenure issues in the long term. While the Philippines has enacted laws to recognize ancestral domain claims, most nations in Southeast Asia have not yet addressed this fundamental land law issue in any depth. Nonetheless, in Cambodia the new Land Law provides for communal title, while Indonesia's constitution respects customary law (*adat*), and claims that stem from it.

According to a CIFOR review of CF policies in India, the Philippines, and China the first generation of laws and policies are badly flawed and "many of the current policies will have to be repealed, replaced, or drastically reformed, [60] as they do not provide adequate authority, livelihood opportunities, or the flexibility for adaptive management to take place. The coming decades will show whether new laws and policies emerge extending broader rights to communities for local forests and land resources, while empowering them with the authority to craft their own systems of management. Further, it is an open question whether national and international leaders will support these policies and laws by ensuring the political and financial support needed to enact this transition.

Legitimate indigenous community claims to lands and forests that are considered "ancestral domain" may be rejected if rural people accept community forestry stewardship agreements that acknowledge the land as state public domain. In Cambodia, an independent forest sector review team recommended that all claims of indigenous groups for communal title be settled prior to the demarcation of state lands and forests.[61] But, no Asian nation has made any significant headway in finding operational methods to integrate customary laws within modern national legal frameworks, even though some have acknowledged the validity of traditional legal frameworks (Indonesia, Northeast India Sixth Schedule areas).

CHAPTER 6

National Program Implementation

Over the last two decades, community forest management has gained attention all over the world as a viable approach to public forest land management. Its growing popularity is reflected in the ratification of CF related laws, the adoption of supportive policies, the expanding investments of bilateral and multilateral agencies in CF programs, the broadening engagement of international, national, and local NGOs and academic institutions in CF activities, and the emergence of community-based forestry networks and associations. As discussed in Part III, a number of Asian countries are developing national policies and laws that formally extend new forest management responsibilities, rights, and roles to communities. Forest policy goals are increasingly moving away from industrial timber supply towards multi-purpose and adaptive management to better serve both the environment and society. Donor agencies are making community participation and livelihood needs a major part of their development agenda and funding priorities. An increasing number of government foresters who once bitterly criticized forest-dependent people as the root cause of deforestation now view them as the best sector agents to restore and protect watersheds and forest lands. Forests and planners who see the value in crafting partnerships with rural villages in order to better manage forests are often confronted by more conservative colleagues who endorse older management styles and by business interests that have profited in the past from conventional systems of state forest control.

While the reorientation of the Asian forestry sector is being guided from above, it is gathering momentum on the ground where a growing number of communities are networking, federating, and politically mobilizing around shared concerns regarding forest and watershed management, livelihood needs, and cultural preservation, supported by a rapidly expanding body of social organizations. Non-government groups involved in generic rural development activities are increasingly engaging in more specialized community forest management support projects, and gaining expertise in this area. Community resource management has also become a popular topic of research and extension for academic institutions, while international organizations are starting to monitor the extent of transition to community forest management. The Food and Agriculture Organization has begun to integrate these types of information into global forest statistics. [62]

According to Forest Trends, at least 22 percent of all forest land in developing countries is owned or accessed by communities.[63] In Asia, it is likely that the majority of the region's forests are accessed by rural communities, since indigenous cultural communities populate and utilize even the most remote lowland forests and upland watersheds. Hamlets, clans, and households allocate forest management rights according to customary practices and laws, and in many areas have been negotiating territorial rights for generations. While community forestry is gaining formal recognition in Asia through a new generation of policies and laws, new programs are generally in an early phase of implementation. As a consequence, the integration of national forest sector policies and informal systems of forest stewardship is limited in scope and quality.

A gross indicator of impact is the proportion of state forest land under community management. The effectiveness of CF efforts may be best reflected in terms of restoring and protecting natural forests and agroforests, as well as in improving the socio-economic conditions of forest-dependent communities. At this point of Asia's forest sector transition, the percentage of all public forests actually managed by communities in a sustainable way with security of tenure and access to markets is likely very limited and difficult to quantify in any verifiable manner.

Community forest management has existed in many areas of Southeast Asia for thousands of years, but the concept of state governments establishing formal management agreements with communities, many of whom still utilize the forests in traditional ways, is largely a new phenomenon. Throughout Southeast Asia, an expanding number of government line agencies are exploring how to best involve communities in resource management and relate to their customary practices including departments of forestry, public works, watershed development, protected areas, and so forth. In some nations and some agencies, this experimentation is being conducted through pilot projects, while other countries are pursuing an accelerated conversion to community management driven by new national policies and programs supported through massive amounts of development assistance. Typically, support from government and technical agencies for forest sector reform is enhanced while project funds are flowing. However, with the termination of program financing conservative positions may resurface within forestry agencies and the political will to implement community forestry may dissipate. In Andhra Pradesh, India, for example, within a year of the conclusion of the first phase of the World Bank joint forest management loan, 90 percent of the 6,000 community forest protection committees were estimated to have ceased to function, though after the initiation of the second phase the village forest groups were reactivated.[64] While the new program attempted to address the heavy dependence on project-based employment as a catalyst for participation, the program continues to struggle under a number of limitations. As a recent assessment notes:

> *While new programs and policies demonstrate a rhetorical commitment to greater devolution and greater clarity in product sharing arrangements, a number of factors constrain the implementation of the new program. These include the interests of powerful individuals involved in timber smuggling, the existence of judicial decisions that restrict legitimate thinking and felling, forest department regulations that control forest product trade, and the ongoing reluctance of the forest department to transfer significant rights to communities to development and implement forest management activities.*[65]

While the laws and policies that support and extend new rights and responsibilities to community managers of state forest land vary widely depending on national contexts, emerging processes to transfer management authority often share many common features. The process of empowering community forest management institutions is being increasingly defined, but it is often confounded by a multiplicity of other rulings and interpretations as well as fast track development schemes. The process, at least in theory, usually requires a dialogue between government representatives and

FIGURE 2: Common Steps to Formalize Community Forest Management Agreements

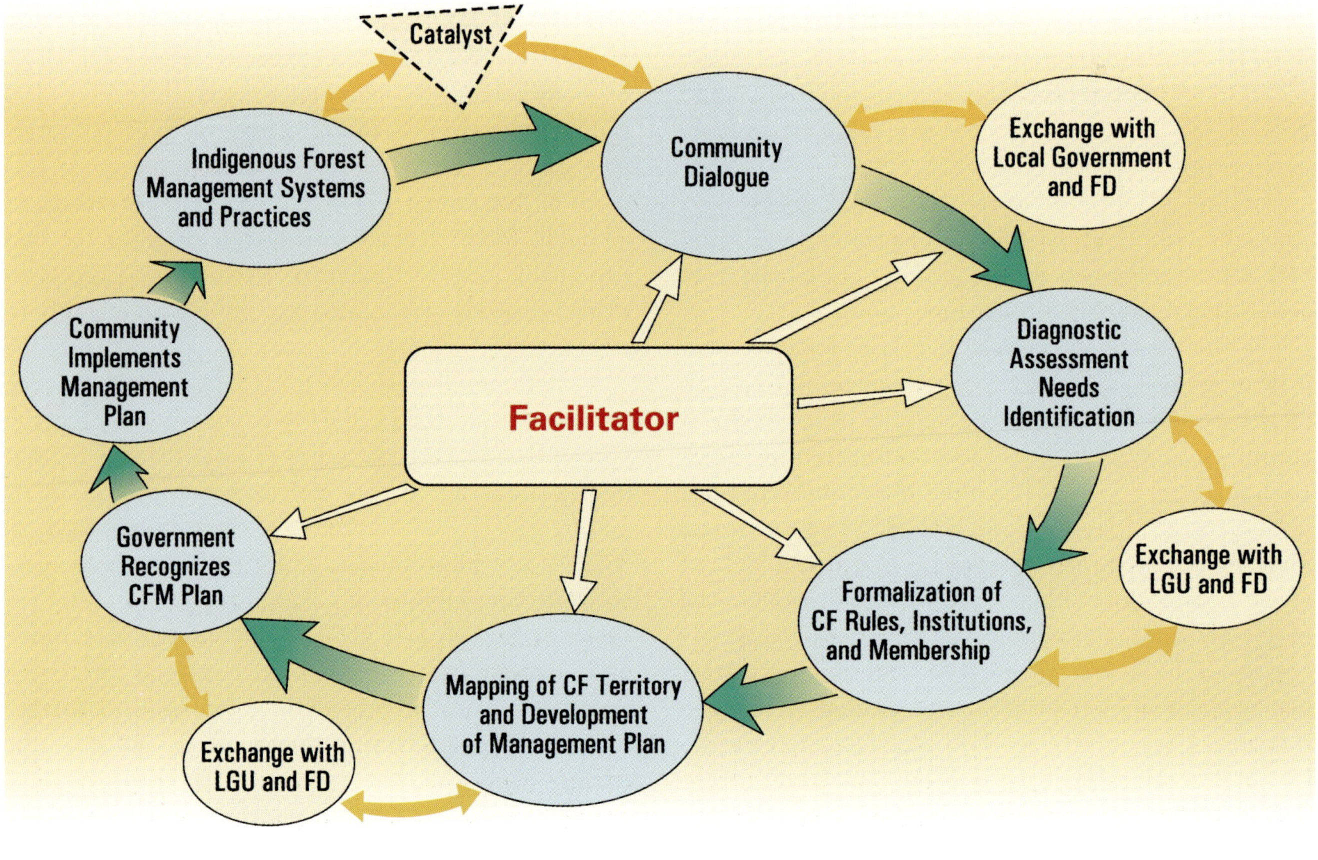

relevant communities to reach a formal agreement regarding management duties, as well as tenure and usufruct rights. The dialogue usually takes place over a period of time, typically from 6 months to several years or more to reach an initial agreement, and continuing throughout the duration of the management contract. At a number of points during the process, key decisions and agreements are made with corresponding outputs. Important elements in this process are depicted in Figure 2. The process and key outputs are increasingly being codified in government orders and implementation guidelines, often accompanying CF legislation and national CF policies. These instructions are sent to local governments and the territorial staff of relevant forestry agencies for implementation. Often the procedures have been under development in government and NGO-sponsored pilot projects for a number of years and are then assimilated into formal procedures from the implementation of new laws and policies.

In many cases, the process begins with communities that already possess informal systems of forest management, either of a traditional nature or of more recent origin. An external or internal catalyst, sometime a forest department official, community leader, or NGO development worker, may initiate a community dialogue about forestry problems and concerns, or opportunities emerging from a new government program or policy. As the dialogue proceeds, local government and other communities may be drawn into the discussions. Most CF dialogue processes include an initial diagnostic assessment to identify management problems and opportunities, informing and setting the agenda for discussions. Community management institutions usually are required to formalize their structure and by-laws, clarifying how executive bodies operate and membership is determined, as well as rules and regulations governing forest use. Defining and mapping CF territory and developing management plans are also important components of government-community dialogue and agreements. Recognition of the management plan is often viewed as the final step in approving community forest utilization rights and responsibilities. The facilitator or assisting organization often may continue to help in the skills development needed in implementing the plan.

Facilitating such processes is a time consuming task and requires skills in community organizing, while communities themselves often need to develop new capacities to formalize their institutions, strengthen and articulate their management mechanisms, and design new management goals and plans. Given the millions of hectares involved and the hundreds of thousands of rural villages in Asia that need to be engaged in such processes, the human and financial resources required to implement such processes on a national level are immense and will require decades to develop.

At present, such constraints are a major impediment to the implementation of new CF laws and policies in Asia and the long-term forest sector transition. For thousands of communities, such constraints frustrate efforts to secure forest and land tenure at a time when external pressures on land are growing in intensity, and delay the start of a generational period necessary for social investment to have effect. In the Philippines, while many CF sites have been recognized by government and while dialogues are ongoing with CF groups, many of the community forest management plans are yet to be completed. A recent review of initially 139 such sites on the island of Bohol showed that the DENR only acknowledges 65 as having agreements under CBFM program and that these plans still need to be internalized by participating communities.[66]

It is very difficult, however, for desperately poor communities to operationalize management plans when there are no resources to assist in the restoration of degraded lands and where basic services are lacking. In many states in India, JFM dialogue processes have often been done hurriedly in order to achieve targets for group formation and finalize micro plans needed to release project funds. Unfortunately, this has not allowed needed community discussions and participatory planning to occur and has undermined the development of community-level management capacity. As a result, the micro plans are often inadequate and do not reflect the real needs and priorities of local forest protection committees. While CF groups are now formally recognized they may not have a strong consensus regarding how to manage state forest land.

As the discussion in Part III suggests, successful implementation of national CF programs requires a convergence of a) empowering legislation, b) political will to implement new laws, and c) human and financial resources to carry out field-level implementation activities. Part IV examines the recent experiences of five Southeast Asian countries in their attempt to develop national legal frameworks and programs that support community forest management.

PART IV

Country Experiences

Community forest management is an important feature of indigenous resource use in much of Southeast Asia. Forest stewardship has evolved over the millennia, becoming increasingly market-oriented in recent centuries as multi-storied forest garden systems with a diversity of tree crops and NTFPs have developed. Most villages seek to achieve multiple goals from their forests including water conservation, micro-climate management, subsistence goods production, recreation, and fuel generation. During the 20th century, indigenous forms of forest management have been under growing pressure, sometimes resulting in a breakdown of community institutions and a loss of their operational effectiveness, while in other contexts external forces are catalyzing community organizations and leaders to strengthen conservation and use controls. This diversity of indigenous and informal forest management practices is now in the process of being formally integrated into the national forestry sectors' policies, laws, and programs across the Asia region. The process is significant, as it potentially affects the lives of hundreds of millions of people and the future of much of the region's forest lands. Each Asian nation is on its own distinctive path towards the recognition of the rights and responsibilities of forest-dependent peoples. As the country reviews indicate, the process has been far from smooth, with significant resistance from powerful interests in government and the private sector. As the following stories show, it is a journey begun, though still not completed. This section presents five country reviews that briefly summarize the processes and actions that are leading to the creation of national community forestry laws and related legislation, policies, and programs that may formally engage communities in the management of the public forest estate. Each country report begins with a discussion of the history of CF related laws and policies, and the processes and forces that have facilitated their establishment, followed by an assessment of the impact of CF policy and program implementation. The national profiles briefly describe the status of community forestry in Cambodia, Indonesia, the Philippines, Thailand and Vietnam, both in terms of the impact of nationally sponsored programs, as well as both traditional and emerging CF groups that operate informally in rural areas. This includes the extent to which community forest management groups are being officially recognized in their nations, as well as the ongoing coverage and importance of informal systems of forest stewardship. Each country review is accompanied by a case study illustrating the ways through which CF systems are evolving. The case study abstracts are drawn from research and pilot activities facilitated under the Community Forest Management Support Program (CFMSP), a four-year program that was funded by the European Commission and USAID and implemented by Asia Forest Network and Community Forestry International to assist local governments and NGOs in implementing CF related policies in five Asian nations.

CHAPTER 7

Cambodia

Cambodia is currently at a historic juncture as it explores alternative approaches to managing state forest lands. It is important to briefly review the national forest policy experience of the past decade in order to understand the potential importance of community forestry in Cambodia's future. Throughout the 1990s, community forestry maintained a low profile in Cambodia's forest policy sector. National policy discussions focused on commercial timber concessions as the primary method for state forest management. In 1993, a national election was held and a coalition government formed. With few financial assets, and limited resources to broker political alliances, the military and political leadership began relying on the leasing of forest logging concessions to generate support and build partnerships. The leadership of the Royal Government of Cambodia justified these policies in the belief that the responsibility for forests could be delegated to the private sector, while generating revenues for national development.

The impact of commercial logging on Cambodia's forests was dramatic. By 1997, 30 companies were licensed to log approximately 6.5 million hectares of forest land, representing more than half the country's forest area.[67] In the mid-1960s, good forest cover, defined as forest with greater than 40 percent crown cover, was believed to extend over 75 percent of the country; however, this had fallen to an estimated 30 percent by the late 1990s. As problems with unsustainable harvesting and illegal logging began to emerge, donors and government planners demanded more sustainable forest management (SFM) that could deliver timber revenues to government on a long-term basis, while conserving the environment and creating local income generating opportunities.

In an effort to improve the sustainability of commercial timber extraction, in 1996 the World Bank and the Asian Development Bank negotiated loans to fund the government to develop a system for timber concession management planning and monitoring. These included the World Bank-funded Forest Policy Reform Project (1997-1998) and the Forest Concession Management and Control Project (2001-2003), while the Asian Development Bank financed the Sustainable Forest Management Project (1999-2000). Both multilateral organizations strongly endorsed concession management as a way to achieve sectoral development objectives and anticipated that a market-oriented policy would lead to revenue of over $100 million per year entering government coffers.[68] Soon after the initiation of the forest reform projects, staff and consultants encountered immense problems with corruption, rampant uncontrolled and illegal logging, and minimal government income from concessions totaling only $6 million in 1997, and $10 million in 1998.[69] Despite initial problems, many of the advisors and supporters of the concession approach continued to believe the obstacles could be overcome, reflected in an Associates for Rural Development analysis from 1998 that notes: "The Forest Concession system is the most appropriate for commercial development of forest resources in Cambodia, but needs refinement to suit Cambodian conditions."[70]

The confidence in SFM, however, continued to erode as field reports and analysis indicated extremely poor performance. An ADB project review published in 2000 noted a "total system failure," with "no concession managed sustainably."[71] In May 2000, the Department of Forestry and Wildlife met with donors and concessionaires and an agreement was reached that no logging would be undertaken until concessionaires submitted acceptable management plans, with a deadline set for September 2001. A donor coalition urged the Government to adopt a policy permitting concession forestry operations only after sustainable management plans had been approved, driven in part, by World Bank loan conditions (known as the Structural Adjustment Credit Action Plan on Forestry). The deadline for SFM plans passed and up to the present time, few concessions have submitted acceptable plans.

In a recent analysis of Cambodia's experiences with concession forestry, drawing on consultant studies, Bruce McKenney concludes that SFM failed to be adopted primarily because it was not profitable to implement practices that would reduce yields from 50 m^3 per hectare to 10 m^3 per hectare. He notes that concessionaires reduced risk by rapid exploitation and attempting to maximize income in a short period of time. He reports that after decades of illegal felling during the period of civil disorder, only 6 percent of Cambodia's forests retain commercially valuable dense evergreen forests, while the remaining evergreen forests covering 30 percent of the country's forest area ranges from moderately to severely disturbed.[72] A study of concession system profitability conducted in 2001 found that "over the past three years (1997-1999) the average Cambodian producer has not generated significant positive net revenues and has never met profit expectations," and there was little, if any, revenue available for government taxes and royalties.[73] A World Bank assessment of concession profitability found that most of the concession areas were not economically viable for timber operations, with 4 million hectares of concessions subsequently cancelled.[74]

McKenney's 2002 assessment makes four recommendations, including: 1) recognize the forest concession model has failed to generate revenues for the state and support rural development, 2) terminate concessions without a viable base of commercial timber, 3) designate forest areas for outright protection where biodiversity, watershed conservation, and eco-tourism are deemed important, and 4) increase resources available for developing and pilot testing community forest management models.[75] While concession-oriented forest management dominated sectoral policies and strategies during the 1990s in Cambodia, a parallel dialogue was beginning to take place that focused on community-oriented management of resources, while new laws were being promulgated that supported the role of communities and local government. The pending termination of logging licenses covering over 4 million hectares of state production forest opens the door for community forestry as a new national management strategy for the forest sector.

National Law and Policy Environment: 1993-2003

While initial forest sector policies under the coalition government focused on commercial timber extraction and large-scale concession management, discussion of community forestry began as early as 1994. The rights of communities to derive subsistence goods from state forest land were articulated in the 1993 Constitution, and in subsequent forest laws. In Cambodia the hierarchy of laws includes the following:

- Constitution
- Laws (*CHBAB*)
- Royal Decrees (*REACH KRET*)
- Sub-Decrees (*ANU-KRET*)
- Implementing Orders (*PRAKAS*)
- Provincial or Commune Regulations (*DEIKA*)

As a whole, a growing body of Cambodia's laws, royal decrees, sub-decrees, and implementing orders provide an impressive body of legal decisions that support the rights of local governments and communities to engage in natural resource management. The Constitution notes the rights of communities to own land and forests privately and collectively, and to form associations. The Land Law recognizes the rights of indigenous communities to communal ownership of land used for residence, agriculture and shifting cultivation. While implementation guidelines for this article are still being developed, it opens the door for the recognition of ancestral domain claims of Cambodia's ethnic communities, a critical issue in the northeastern part of the country where indigenous people's traditional territory is under growing land pressure. The recent proliferation of Cambodian laws recognizing community rights, responsibilities, and roles in administering natural resources are presented in Tables 4 and 5.

The Commune Law enables Commune Councils to assess resource use and needs of local people in preparing development plans, including those for natural resources, while the Environmental Protection Law ensures that people have the right to access information and participate in environmental protection and natural resource management. For rural Cambodians dependent on natural resources for their livelihoods, the Forest Law ensures a number of important rights to subsistence, forest-based activities including the right to gather NTFPs for customary use, to collect resin, to graze livestock, and to harvest timber for houses and fences. While a legal endorsement of subsistence use rights is reassuring for forest-dependent people, communities that are interested in small-scale commercial utilization may be confronted by a range of harvest, transit and trade regulations.

Since the Forest Law covers a broad range of issues, a Community Forestry Sub-Decree was drafted to provide greater detail concerning the rights and responsibilities of communities engaged in collaborative management with the Forest Administration. The CF Sub-Decree allows communities to request a community forest agreement in order to manage an area

TABLE 4: Natural Resource Management Rights of Communities Under Cambodian Law

Topic	LEGAL RIGHTS	LEGAL RESPONSIBILITIES
Land Based Tenure	• Citizens have the right to own land either individually or collectively (C 44) • Local communities may access forest concessions to gather forest products (FL 15, FCM 6) • Local communities have the right to establish an area of forest under a community forest agreement for a 15 year term (CFSD, FL 43) • Indigenous communities are entitled to collective ownership of land used for residence, agriculture & shifting cultivation (LL 24) • Concessionaires must respect community title and remove area from the concession (FL 15)	• State owns the forest resources with the control, use and management to be determined by law (C 58) • Indigenous communities may not sell or transfer land under collective ownership (LL26, 28) • People may no longer acquire ownership of land through clearing forests and possessing land for five years (LL 30) • It is prohibited to clear forests for a road or to settle (FL 31) • MAFF shall coordinate with communities on surveying forest land and establishing collective ownership (FL 10)
Natural Resource Use	• Right to gather NTFPs for customary use (FL 24, 40) • Right to harvest timber for houses, fences, etc. (FL 24, 40) • Right to barter and sell NTFPs at sustainable level (FL 40) • Right to maintain, harvest and gain profit from trees planted by community or individuals (FL 46) • A community under a community forest agreement has the right to commercially harvest timber five years after approval of management plan (CFSD)	• Commune Councils shall assess resource use and needs of local people in preparing development plans (CL 63) • Communities not under a community forest agreement require a permit to use NTFPs above customary use (FL 26) • Prohibited to harm resin tree tapped by local people (FL 29) • Prohibited to use chain saw in forest without permit (FL 38) • DFW shall prioritize giving permits to local communities to gather forest resources from within a concession (FCM 6)
Natural Resource Management and Conservation	• Community may request to form community forest agreement for an area of forest for 15 years (CF) • Communities may request technical assistance from any qualified individual, civil organization, NGO or govt. authority (CFSD) • Communities under a community forest agreement shall operate under a management plan to be reviewed every 5 years (CFSD) • The government shall respect and assist local communities in designation and protection of the spirit forest (CFSD, FL 45)	• The Commune Council shall manage resources in sustainable manner in meeting needs of local people (CL 41) • The Commune Council shall conserve natural resources and environment in promoting economic development (CL 43) • Resources under a CFA must be used in a sustainable manner (CFSD) • A land concession shall only be granted on State Private Land and should not include natural forest (LL 58)

of forest for 15 years, with required management plans being reviewed every 5 years. The Sub-Decree defines a community as "a group of residents in one or more villages in the Kingdom of Cambodia who share a common social, cultural, traditional and economic interest and use the natural resources in an area, where they live in or near, in a sustainable way for subsistence and livelihood improvement purposes."

The agreement may be renewed after the 15-year period if the area is being managed in a sustainable manner in compliance with the agreement, approved management plan and rules and regulations related to community forestry. Communities are also allowed to request technical assistance from any qualified individual, civic organization, NGO or government authority to establish community forestry systems. Finally both the Forest Law and the CF Sub-Decree require the government to respect and assist local communities in designation and the protection of spirit forests. In addition to the legal framework described above, the Cambodian Government and international agencies have taken a number of important policy decisions in recent years that support decentralization and community engagement in natural resource management.

Since the issuance and approval of the CF Sub-Decree, the Forest Administration, in conjunction with donor agencies and NGOs, has been engaged in drafting implementation guidelines (*prakas*) that define the operational steps communities must take in order to secure a forest management agreement and gain approval of their management plan. The implementation

TABLE 5: Cambodian Policies Supporting Community-Based Natural Resource Management

Date	Policy Action	Approving Official
Early 2000	Release of 8,000 hectares of commercial fisheries to communities	Prime Minister
Early 2001	Release of 536,000 hectares of commercial fisheries to communities (56 percent of total commercial fishery area)	Prime Minister
Late 2002	Loan from Asian Development Bank for Community Oriented NRM project around Tonle Sap	Royal Government of Cambodia
2003	Cancellation of 4 million hectares of commercial timber concessions (over 60 percent of total area leased to loggers)	Prime Minister
2002	Approval of 14,000 hectares of O Pa forest for Wat and Community Management in Oddar Meanchey	Prime Minister and Governor of Oddar Meanchey Province
1998-2003	Approval of 50,000 hectares of forest for Community Management in Ratanakiri	Governor of Ratanakiri Province
1998-2003	Approval of 110,000 hectares of flood and upland forest for Community Management in Siem Reap Province	Governor of Siem Reap Province

guidelines were necessary to clarify what the FA required from the community in terms of 1) CF group organization and by-laws, 2) forest rules and regulations, 3) procedures for securing forest agreements, and 4) guidelines for the preparation and approval of forest management plans.

On June 1, 2001, the Prime Minister turned over 56 percent of Cambodia's commercially zoned fishing concessions to community management covering 536,000 hectares of fishing lots.[76] The Government recently finalized a loan from the Asian Development Bank to finance community fisheries management around the Tonle Sap where a significant area of flood forest needs management. The suspension of 4 million hectares of forest concession creates greatly expanded opportunities for community forest stewardship in the uplands. Finally, Governors in a number of Cambodian provinces give recognition to the efforts of communities, monastic groups, and indigenous people to manage local resources (Table 5). While the impact of these policies is yet to be determined, they indicate an orientation of the political leadership towards a greater decentralization of authority for natural resource management and a growing political will to shift control of natural assets from private sector operators to local government and community leadership.

Process for Formulating the Community Forestry Sub-Decree

The formulation of a legal instrument to formalize community forestry in Cambodia has evolved over the past decade. While forest concessions dominated Cambodia's forestry sector throughout the 1990s, community forestry initiatives developed gradually during the same period with far less publicity. Externally aided community forestry projects began emerging soon after Cambodia started receiving international development assistance in the early 1990s. In 1992, the Mennonite Central Committee began supporting pilot projects in Takeo Province, and soon after Concern Worldwide initiated CF field activities in Kampong Chanang and Pursat Province. Gradually other international NGOs and Cambodian NGOs began experimenting with a wide variety of approaches to community forestry. FAO supported the first provincial strategy for community-based natural resources management in Siem Reap in 1995, a program that grew in coverage as it increased its emphasis on participatory fisheries and forest management over the next ten years. Throughout the 1990s, a number of staff of the Department of Forestry and Wildlife (DFW) frequently participated in national, regional, and international meetings on CF subjects. By 1994, some planners within the DFW were already beginning to explore possible policy and legal frameworks for CF.

Throughout the mid-1990s, interagency meetings were held to share early field experiences with CF among practitioners and by 1998 national working groups and networks were being established to explore national approaches to community forest management in Cambodia. In 1998, DFW established a Community Forestry Unit and began building a national CF data base, aided by support from GTZ's Sustainable Management of Natural Resources (SMRP) Project and later by the Cambodian German Forestry Project

(CGFP) (Table 6). The national working group soon evolved into a CF Strategic Planning Team, comprised of individuals from DFW, GTZ, JICA, ADB, Concern Worldwide, IDRC, and FAO, and issued the first National Community Forestry Strategic Plan in 2000.[77]

With a growing consensus forming around the National CF Strategic Plan, the CF Strategic Planning Team turned its attention to the need for a legal framework to enable CF to take place formally. In 2001, a Task Force was established to develop a draft CF Sub-Decree (*Anukret)* that could be distributed for public comment. A six-month consultation, drafting, and comment period was held to gain input from stakeholders around Cambodia between August 2001 and February 2002. Upon completion of a final version in February of 2002, the drafting team submitted the document to DFW for internal review prior to submission to the Ministry of Agriculture, Forestry and Fisheries (MAFF) and ultimately the Council of Ministers.

From February through June 2002 a continuing dialogue took place between the DFW and members of the Task Force in an effort to agree on key elements within the CF Sub-Decree. Aside from dealing with such issues as benefit sharing, CF management plan approval procedures, and committee structures, a key goal of the Task Force during the entire process was to ensure that the draft Sub Decree (*Anukret*) was informed by a broad group of stakeholders and that there was a strong sense of ownership for the CF Sub-Decree, both within government and among donors and NGOs.[78] Part of the success of the consultation process was the presence of several NGO networks and forums that facilitated input from their members, together with the ability of government and non-government members of the Task Force to work collaboratively.

For over a year, the Community Forestry Sub-Decree was reviewed and revised by the DFW as comments were received from key stakeholders, MAFF and the Council of Ministers. In late 2003, the CF Sub-Decree was finally approved. Since many operational guidelines were eliminated from the final draft, these are now being formulated for issuance as Implementing Guidelines (*Prakas).* The *Prakas* will cover four main areas including: 1) Management Committee organization, 2) Organizational By-Laws, 3) Management Plan, and 4) Agreement. The draft orders were circulated among stakeholders around the third quarter of 2004.

The DF&W was reorganized in early 2004, adopting a new organizational structure and renamed the Forest Administration (FA). The Community Forest Unit has been upgraded to an Office and the development of a National CF Program is currently under way, drawing on input from four regional workshops and a national synthesis meeting planned for the end of the year. The FA is attempting to develop a road map for national CF implementation that it hopes will attract international financing from development agencies.

Prospects for Policy Implementation

Both the policy and legal environment in Cambodia has become much more supportive of community forest management over the past five years

TABLE 6: Community Forestry Policy Events In Cambodia (1992-2004)

1992	• Community Forestry Projects initiated by international NGOs (Concern, MCC, and FAO) and Department of Forests and Wildlife (DF&W)
1994	• DF&W Drafts Community Forestry Sub-Decree (CFSD)
1996	• 1st National CF Network Meeting
1998	• Community Forestry Working Group established by DF&W and MRC/GTZ
2000	• National Community Forestry Strategic Plan (final draft issued by DF&W and • Developed by the Community Forestry Strategic Planning Team)
2001	• CF Task Force formed to accelerate development and approval of the CF Sub-Decree
2002	• CFSD Public Hearings, Task Force Meetings, Comments and Revisions – Final Draft of CFSD submitted to government
2003	• CFSD Approved by Council of Ministers and signed by the Prime Minister
2004	• Development of National CF Program – Restructuring of DFW to Become Forest Administration (FA) • Formulation of CF Implementation Guidelines (Prakas) • Training of FA staff

(2000-2004). The suspension of 4 million hectares of timber concession that have failed to pass economic viability reviews has created potential opportunities for new forms of management to take their place. This may result from emerging shifts in national forest management policies. Some provincial and district governors, as well as commune councils, are taking greater interest in the state of forests in their administrative territories, especially where military or private sector operators have used the resources unsustainably and without benefit to local government. Some local government administrations view local communities as natural allies in efforts to help stabilize forests and threatened watersheds.

The emergence of the CF Sub-Decree and forthcoming implementation orders (*Prakas*) provide legally recognized rights and responsibilities, and specific procedures communities can follow to gain formal management authority over local forests. The Forest Administration has promoted the formalization of community forestry and the development of a national CF program and is currently seeking financing to initiate a large-scale program. Donors are also looking for new approaches to funding the Cambodian forest sector. As part of this effort, a Forest Sector Review was financed by the Joint Coordinating Committee, consisting of 6 ministries and 6 donors, under the chairmanship of the Director of the Forest Administration and the Chairman of the Working Group on Natural Resource Management.[79] This review recommended the termination of industrial logging under large private concessions, and the devolution of forest management to local governments and communities.

Now that a more supportive policy and legal framework is emerging in Cambodia, the question is what will be the extent of implementation of the new policies and laws? There remain influential vested interests that support the continuation of commercial timber felling and land clearing. In 2004 and 2005, the incidence of illegal land grabs rose sharply, especially in areas with potential for cash crop development. Efforts of "powermen," as they are locally referred to, to manipulate local government and secure community lands were so blatant and widespread that Prime Minister Hun Sen made a public statement banning illegal land sales. Nonetheless, there continues to be a tension in Cambodia between newly passed land, resource, and governance laws that endorse many community rights and demand a transparent process, and rampant, behind-the-scenes, resource and land speculation.

Funding and capacity to facilitate a national transition in forest management systems also represent a major constraint. Nonetheless, Cambodia has made considerable progress in crafting an enabling policy and legal environment in which CF systems can expand. The challenge at the present time is developing a national capacity to implement the new laws.

National CF Program Implementation

Rural Khmer communities, as well as other indigenous peoples living in Cambodia, have played an important role in the management of Cambodia's forests for centuries. In some places, indigenous communities retain ancestral claims to forest lands and possess indigenous methods for management decision-making. Often such lands are located in areas with traditions of swidden cultivation (*chamkar*), where land and forest resources have been communally owned. This is particularly common in Northeastern Cambodia. In other areas, forests have been important sources of income through the collection of non-timber forest products including resins, rattan, oils, honey, aromatic wood, bird's nests, mushrooms, insects, and other goods. NTFPs like these were among the primary export commodities for at least the past thousand years. One of the most important forest management activities has involved the tapping of yang oil from the *Dipterocarpus alatus* tree that mainly grows in mature semi-evergreen or evergreen forests. Yang oil trees are usually marked and held under the ownership of individual households according to customary law.[80] To that extent, the forest may be communally owned and managed, while individual trees may be privately held.

In Cambodia, traditional systems of forest stewardship have been under pressure since the advent of the French colonial policies covering state forest management. Indigenous forms of community resource management were further disrupted during the extended period of civil disruption in the 1970s, '80s, and early '90s. While indigenous forms of community forestry have not been widely documented, to underestimate their presence and informal role in forest management denies the importance of their social capital that could play a strategic role in stabilizing forest ecosystems in many parts of the country. Equally importantly, there is a growing movement in many rural communities, especially those experiencing increased pressures on their forests, to establish new systems of forest protection. There is an expanding body of reports from field visits that indicate communities are concerned about illegal logging in forests neighboring their communities and increasingly they are organizing committees in order to protect them and coordinate with local government and the Forest Administration. Since the CF Sub-Decree has only recently been approved and implementation guidelines (*prakas*) are still under development, only a small minority of existing CF groups have received formal recognition.

In examining the impact of CF on the ground in Cambodia, it is important to differentiate major contexts within the country. Much of the early CF efforts assisted by external organizations took place in the densely populated lower Mekong region, where very few natural forests exist. In that context, most community forestry efforts focused on plantation establishment and regeneration of small patches of degraded forests. A second major context was the flood forests around Tonle Sap, while a third setting was the mangrove forests along the southern coast, where a number of donors were engaged. The larger tracts of natural forests, however, are located in the Upper Mekong and the Northeast, the Northwest, and the Cardamom Mountains. These areas have generally received less assistance in the form of CF programs, but frequently are inhabited by Khmer and indigenous peoples with long standing traditions of forest management. This diversity in Cambodia has required national planners, donors, and NGOs to adopt different approaches as they attempt to support CF through policies and extension programs.

Official Coverage of CF

The most comprehensive study to date concerning the impact of community forestry in Cambodia was conducted by the Cambodian-German Forestry Project in November 2002.[81] The study focused primarily on 57 externally supported projects in 15 provinces. The report identified 83,000 hectares under CF management included mangrove forests, totaling 0.7 percent of the nation's forest area and included 404 villages, representing 3.6 percent of the population.[82] The assessment also suggests that if one assumed communities could protect forests within a 10-km walking distance of their settlement, 8.4 million hectares of Cambodian forest could be brought under community management, with an additional 3.9 million hectares of agricultural land converted to farm forestry. Should this projection be achieved it would bring the majority of Cambodia's 10.6 million hectares that were classified as forests under legally recognized community stewardship.

If CF has so much potential in Cambodia, why has so little been done to date in terms of formalization? There are a number of reasons for the gap between the actual and potential area under formally recognized systems of community forest management in Cambodia. Over the past ten years, with the exception of the FAO-funded project in Siem Reap, most community forestry has been implemented by smaller NGOs with limited resources. Until the Community Forestry Sub-Decree was approved in October 2003, no formal government legislation supported the role of communities in state forest lands management, and even that process awaits the approval of implementation guidelines.

Led by small NGOs operating primarily in the Lower Mekong region and Ratanakiri province, and by the FAO-supported community fisheries and forestry project in Siem Reap, considerable experience has been acquired regarding the important roles communities can play in stabilizing and restoring Cambodia's forests. There is an increasing number of Cambodians who have experience in CF field practice and policy development who are working with NGOs, international organizations, bilaterally funded projects, and government programs, providing a basis for a national CF program that will accelerate forest sector transitions in the future.

The experience of Cambodia over the past decade is not unlike that of many nations in Southeast Asia. Indigenous systems of community-based forestry have been present in Asia for centuries, and are frequently rediscovered and developed first by NGOs, usually without government endorsement. As the value of community resource management is recognized through the pilot programs sponsored by NGOs and donor projects, the attention of government and larger development agencies is attracted to this approach. A mutually reinforcing process of supportive policies and larger financing often results in the evolution of community forestry from isolated pilot projects to national strategies. Cambodia appears to be moving

MAP 1 : Potential Zones for Community Forest Management
Ratanakiri Province, Cambodia

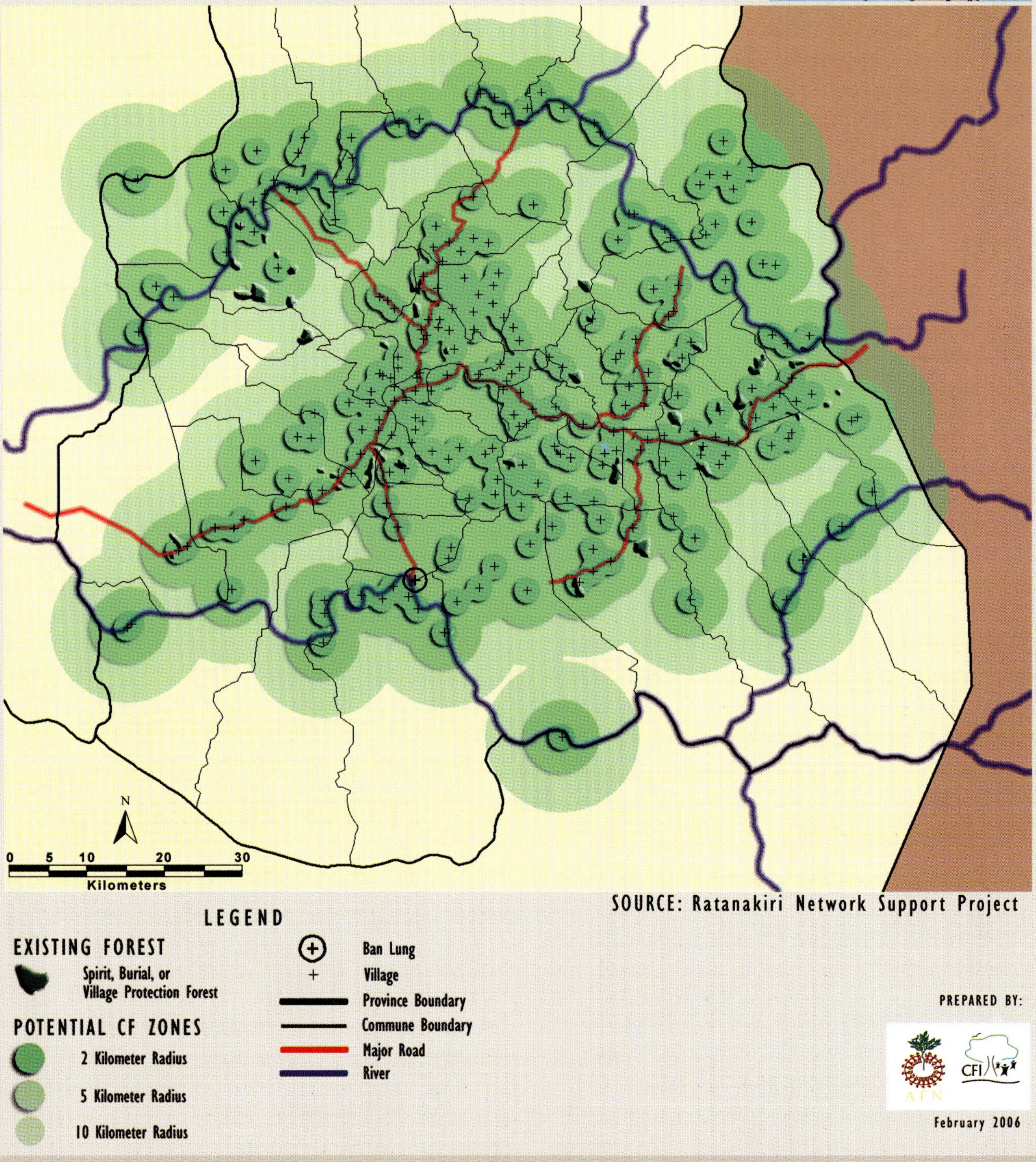

into such a transitional period in the forestry sector. The expansion of community forestry is also linked in Asia to the decline in industrial timber exploitation on state lands, usually as commercially valuable stands are depleted. With the revoking of logging permits, much of this area requires new forms of management. Given the limitations of the Forest Administration's field capacity at the present time, community forestry presents an attractive option.

Informal Coverage of CF

Assessing informal coverage of community forestry systems in Cambodia and their impact on protecting forest lands is a difficult task. Even anecdotal information is limited, as much of the country is isolated from roads. In fact, it is in some of the most isolated regions where informal CF systems are most active. Some indications of the types of informal CF systems operating are being generated through field visits of Community Forestry International (CFI) that made trips to Oddar Meanchey in the northwest and to Kratie Province in the Upper Mekong in the first half of 2004. The CFI team found that one group of monks in Oddar Meanchey had been successfully protecting 14,000 hectares of dry deciduous forests for several years, while illegal logging was occurring in neighboring areas. They also found that a number of villages were attempting to establish forest protection systems with the assistance of local NGOs. In Kratie Province, the team met with four local NGOs that were engaged with 12 villages distributed across the territory, many of them neighboring, or within, timber concession areas. The areas under community management ranged widely in size from 600 hectares to 10,000 hectares or more. An important concern of communities in Kratie province is to protect local forests from commercial loggers, especially the valuable yang oil trees. While only one CF group has received formal recognition from the local Forest Administration, communities in Kratie may be providing protection and management to 50,000 hectares or more on an informal basis.

An assessment conducted in mid-2005 by Community Forestry International found that over one million hectares of forest was under village management belonging to 511 communes.[83] This data indicates that rights are already being exerted by one-third of the nation's commune to 10 percent of the forest area, even prior to the initiation of a national CF program.

In Ratanakiri Province, forest management has been practiced for generations following customary laws and traditions of a number of cultural communities. The half-dozen indigenous populations that reside in the area have used the forests as a pool of land for long-rotation agriculture. Indigenous people have dozens of classifications for their forests including sacred groves, bamboo forests, watershed and spring forests, as well as a large body of forest land that may be cleared periodically for agriculture, used for several years, and then left to regenerate. While land registration is just beginning to take place, and communal tenure has recently been recognized, there is still no official data on the amount of land under communal stewardship. Nonetheless, field reports indicate that most of the forest land in Ratanakiri Province, and probably Mondulkiri as well, falls under the authority of indigenous communities who coordinate with neighboring communities, make decisions on forest protection and utilization, fire control, and attempt to maintain forest cover in the face of growing pressures from outside land speculators. A recent decision by the Office of the President, declaring a moratorium on land sales until land registration has been completed may help reduce the recent increase in land grabs in the northeastern part of Cambodia.

Most recently, a zoning exercise conducted by Community Forestry International for Ratanakiri Province indicates that much of the province could be zoned for community forestry if forests within a 2-, 5- and 10-kilometer radius of each village were allocated for communal ownership or placed under CF agreements (Map 1).[84] Under such a scenario, community stewardship of the provinces' natural resources would be the primary instrument for managing economic growth based on agroforestry and forest products, combined with cultural and conservation-oriented tourism.

If the Royal Government of Cambodia were to adopt a policy similar to that in the Philippines, where ancestral domain claims were rapidly recognized and officially documented, the amount of land under community forest management in northeast Cambodia could exceed one million hectares. Whether such action will be taken will depend on how relevant ministries interpret recent legislation, especially the Land Law, Forest Law, and Community Forestry Sub-Decree.

Cambodia is currently one of the poorest nations in Southeast Asia with limited government resources to

BOX 1: Flood Forests, Fish, and Fishing Villages around Tonle Sap, Cambodia

Kompong Phluk represents an excellent example of the ways some fishing communities residing in the flood forests of Tonle Sap are mobilizing to protect and manage forests and fisheries. This case study of a Community Forestry program is a joint effort involving three hamlets within the commune, the Provincial Departments of Forests and Fisheries, and an FAO project entitled "Participatory Natural Resource Management in the Tonle Sap Region." The purpose of the collaboration is to develop and ratify village-based resource management systems in Siem Reap Province.

Tonle Sap is the heart of mainland Southeast Asia, fed by the mighty Mekong River, sheltered by vast tropical forests that have covered the surrounding hills and plains for millennia. It has played a central role in the evolution of human civilization and continues to be a key factor in the Cambodian economy. In 2000, the estimated value of the annual fish catch of over one million tons was US$ 800 million, representing a major food source for the people of Cambodia, as well as neighboring Thailand and Vietnam. Studies indicate over 1200 species of fish inhabit the Mekong River and the Great Lake, including the 300-kilogram giant freshwater catfish (*Pangasia-nodon gigas*) and the freshwater Irrawaddy dolphin[80].

Forests play a critical role in sustaining the aquatic ecology of the lake. In the upper watersheds, temperate montane and tropical rainforests slow water run-off and greatly reduce erosion and down-stream sedimentation. Around the lake, flood forests protect the lake core during the dry season, and act as an immense hatchery during the rainy season. The health of the Great Lake is also closely linked with the flow of the Mekong River. During the wet season, the lake expands to 5 times its dry season size, and its average depth increases 6 to 8 times, with an estimated 60 percent of the additional water, flowing down the Mekong River, then turning north and reversing its flow into the Tonle Sap River. Ecologically and economically, the Tonle Sap and the Mekong River are among the most important hydrological systems in the world, yet their ability to con-

Kompong Phluk's main street in the wet season is fully flooded, with all traffic moving on boats and barges. Each house becomes an island. The waters of the Tonle Sap begin to rise in June and reach full flood in early October.

Kompong Phluk's main street in the dry season after the lake's waters have receded.

MAP 2. Community Fishing Lots, Tonle Sap, Cambodia

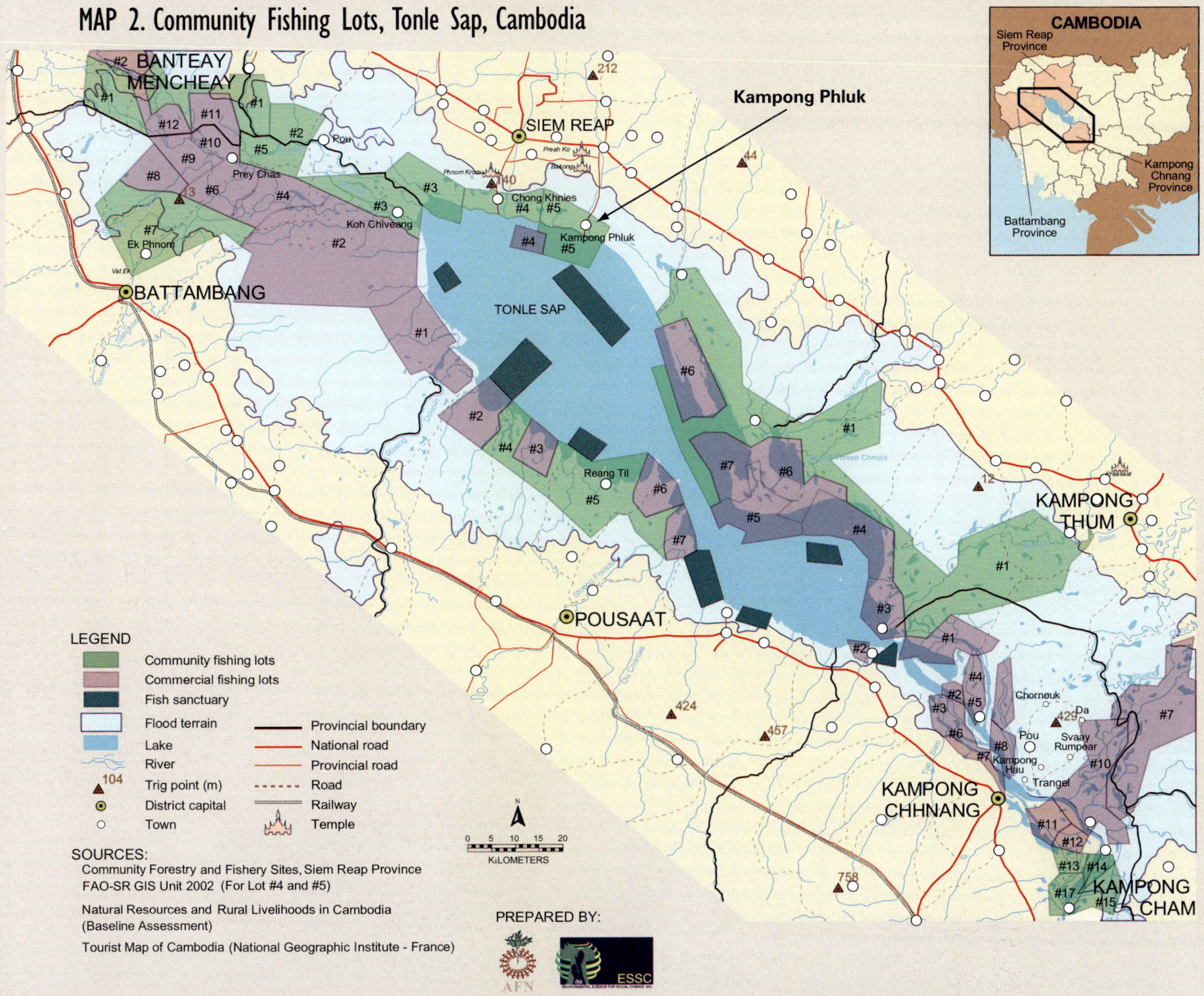

tinue to function as they have in the past is in question, in part due to upstream dam construction, especially in China.

In 2001, the commercial fishing lots established under the French administration were revised and around 20 were given to communities in the flood zone. While eight fish sanctuaries were established (Map 2).

Forest protection is not new to Kompong Phluk. The village first initiated a ban on flood forest cutting in 1944, after clearing for watermelon cultivation threatened the forests that had protected the village from high winds and waves that blew off the lake in the rainy season. Pressures on the forest emerged again as commercial mung bean farmers from surrounding areas sought to clear local flood forests. While the community has been aware of the need to protect the forests for over half a century, the need to strengthen their efforts through an alliance with local government became clear in the late 1990s.

Working with the project support team, village leaders formed a forest and fisheries management committee, elaborated and wrote down rules and regulations for resource use, and developed a management plan for their 5000 hectare area. The management plan included setting sustainable harvest levels for fuelwood, creating protected forest areas, identifying permitted and banned fishing gear, and initiating income generating activities. While external pressures on village resources remain, neighboring communities increasingly accept the management system being implemented in Kompong Phluk.

Nearly 100 other communities in Siem Reap, both those along the lake, as well as in upland watersheds have been assisted by the project to establish forest management systems. An important element in the FAO project was its effort to link communities and local government offices of fisheries and forestry. The initiative demonstrated that Cambodia provincial technical agencies can play a strategic role in facilitating the development of community resource management systems provided they are given assistance in building their extension capacity, both in terms of funding and technical assistance.

The FAO supported project in Siem Reap is also significant as it developed operational strategies to implement new policy and legal reforms taking place in Cambodia that addressed commercial inland fisheries. In Siem Reap Province, the pressures associated with the fishing concession system have helped catalyze the release of 536,000 hectares of commercial fishing grounds, over 50 percent of all fishing lease areas in the country, to the communities for community fisheries management. The transfer occurred on 31 May 2001. In anticipation, the Prime Minister decided to remove all the fisheries staff from the Tonle Sap and sent them back to their offices, creating an open access situation that severely depleted the resources of the lake that year. The fisheries staff are now back in Tonle Sap and slowly, they are learning to work with a growing number of Community Fishery Organizations with support from local and international NGOs.

Rising population pressures in rural Cambodia, rapid policy and political changes, and the introduction of new technologies and markets has overwhelmed traditional systems of flood forest and fisheries management. Overextraction has drawn down fish populations to a point that threatens many species, while undermining the livelihood of thousands of communities that reside around the lake and along the river. The most serious long-term threats to these natural resources, and the populations that depend on them, are deforestation and dam construction in upstream catchments. Deforestation in both upper watersheds and lower watersheds near the lake has occurred at a rapid pace, especially in the past decade, sending enormous loads of silt clogging the river and lake. Commercial logging in China, Burma, Laos, and Cambodia has degraded forests in many parts of the regions that drain into the Mekong River, which in turn feeds the Great Lake. Clearing of over 50 percent of the flood forest that surrounds Tonle Sap, has reduced the riparian buffer that limits the influx of sediment, and substantially lessened fish breeding grounds. ■

Fishing is a primary source of cash income for Kompong Phluk's families. Over 500 species of fish have been identified in the lower Mekong Basin. Dozens of species are caught in the village utilizing a wide-range of dipnets, gillnets, castnets, lines, traps, spears, and other techniques.

support rural development and natural resource management programs. The institutional infrastructure is weak, both in terms of local government as well as the capacity of the Forest Administration at the field level. Most cantonments have few triage staff (field level), especially those with experience in coordinating with commune councils and communities. Commune councils are also new institutions with little experience in resource management. While CF is certainly subject to capture by more powerful groups so are all local government institutions, but with proper capacity building, support, and monitoring, CF still provides much greater opportunities for equitable division of forest benefits among rural populations, than control by private sector firms and government agencies. Over time, as field management activities grow, CF committees can be coordinated through Commune Councils. Initially, however, the emphasis will likely be placed on building resource management capacity at the hamlet and village level, with coordination and networking facilitated through Commune Councils and District and Provincial level government.

While considerable capacity building will be required, a recently completed ten-year project focusing on community-based forestry and fishery management in Siem Reap province (1996-2005) has demonstrated that communities, local government, and technical agencies can work effectively together. Nearly 100 other communities in Siem Reap, both those along the lake, as well as in upland watersheds have been assisted by the project to establish forest management systems. An important element in the FAO project was its effort to link communities and local government offices of fisheries and forestry. The initiative demonstrated that Cambodia provincial technical agencies can play a strategic role in facilitating the development of community resource management systems provided they are given assistance in building their extension capacity, both in terms of funding and technical assistance (Box 1[85]).

As the Kompong Phluk case study illustrates, there is increasing recognition by communities living along the Mekong River and around the Great Lake that these resources are under growing pressure and require careful management. Fortunately, the Government of Cambodia has recently passed new policies and laws to allow villages more rights and responsibilities to protect and manage these resources. While establishing more sustainable systems of community forest and fishery management alone cannot entirely halt the long-term changes impacting the Great Lake, they can make an important contribution to the environment. Strategic actions are needed to build national capacity to develop partnerships between communities, local government, and the Forest Administration (FA). The emergence of national CF Networks, forums, national workshops, and other mechanisms and activities is indicative of the expanding efforts to mobilize government and NGO staff to engage in CF programs. Such trends are gaining momentum as CF is endorsed as a national forest management system by the FA and gains greater legal, policy, and political backing.

Developing a national CF movement to formally transfer management of large areas of state lands to communities represents a complex, large-scale, and lengthy process. Public support and positive attitudes on the part of government planners, political leaders, and local government officials are key components in building the political will to facilitate such a major shift. To establish a viable national policy, it is important to create a broad and shared vision of the future management systems and the societal goals it is expected to achieve. Given the diversity of Cambodia, the implementation design must be sensitive and able to respond to a variety of social and natural environments where it will eventually be implemented. The visioning process can be used to set medium- and long-term goals for the forestry sector, including objectives for responding to rural livelihood requirements, conserving biodiversity, and ensuring important hydrological services.

There are a number of factors that are creating opportunities that facilitate national transitions towards community forestry. As demographic expansion takes place, private sector operations grow, and land requirements increase; forests near communities come under intensifying pressures. Many rural Cambodians are concerned by external demands on local forests and are mobilizing to find ways to control outside use, especially when it threatens their existence. NGOs in Phnom Penh are receiving a growing number of requests from communities to assist them to protect their forests again illegal logging.

Equally encouraging is the shift in forest policy occurring in Cambodia over the past five years with the Royal Government of Cambodia. In 2003 the Cambodian Prime Minister Hun Sen voiced regret over the destruction of the country's forests following criticism from the United Nations of massive logging. "I very much regret the wrong decisions made by the Cambodian Development Council, involving forestry concessions to companies which were issued in large numbers and at cheap prices." [86] Reflecting this change in forest priorities by the national leadership, in 2002 Cambodia put in place new laws to preserve more than one million hectares, or 2.47 million acres, of wilderness in the Cardamom Mountains in the western part of the country, making it the largest wildlife sanctuary in Southeast Asia, while placing a moratorium on most commercial logging operations. By 2005, however, pressure from the private sector to reinstate timber concession activities, at least in some areas, has grown.

A growing number of development agencies have expressed interest in exploring future investments in CF, including the Belgian Government, USAID, the World Bank, JICA, New Zealand, and Australian Aid. The Asian Development Bank has earmarked a substantial proportion of its investment for the Tonle Sap to facilitate community-based management of flood forests and fisheries. There appears to be an increasing number of government staff, donors, and NGOs that believe CF should be a major component of future national forest management strategies, reflected in the formulation of policies, programs, and field based initiatives. As the recent Forest Sector Review notes, "Community forestry has demonstrated that local people are concerned about forests and are prepared to organize around forest protection and management."[87]

There has been a gradual expansion of CF in many provinces around Cambodia led by a growing number of grassroots level NGOs, Buddhist monasteries, and local leaders. The emergence of civil society organizations and processes in Cambodia that are guiding and facilitating a national transition to community forest management also include the CF Network, the NGO Forum's forestry sector group, Concern Worldwide, Oxfam G.B., and Community Forestry International. The efforts of these NGOs, many supported by multilateral and bilateral development agencies, are all contributing to national momentum for change in the forest sector. Perhaps more importantly, community forestry networks are emerging in a number of provinces, and the number of grassroots NGOs and Buddhist Pagodas involved with CF support activities is increasing rapidly.

While there has been a clear trend towards giving greater emphasis to CF in Cambodia, many constraints remain. Despite the suspension of millions of hectares of forest concession, illegal logging remains rampant in many parts of the country. While national policies condemn such destructive behavior, there are strong vested economic interests in continuing illicit timber felling and exports. Little funding exists within the FA to support a CF national program, and no major forestry sector loans are operating or under development to finance community forestry, except for ADB assistance around the Tonle Sap. The vast majority of FA field staff has little training in community forestry, few financial or technical resources to draw on, and little incentive to support community forest protection efforts.

The legal and policy environment emerging in Cambodia presents both supportive and constraining elements for CF. The Forest Law and CF Sub-Decree formally acknowledge the role communities can play in forest management, specifying rights and responsibilities. This is an important step forward in recognizing community resource rights by government. The Land Law even provides an opening for communal title in areas with indigenous people. The recently drafted CF implementation guidelines describe in considerable detail the actions rural communities must take to receive formal recognition of their community forest areas. The limitation of the new legislation is that the terms for community participation often reflect the bureaucratic procedures of the FA and its need to retain significant control over resource utilization, complicating the interface with rural communities that may approach forest management differently. In other words, it may be overly rigid and specific vis-à-vis the terms of engagement and may not provide sufficient utilization incentives. Detail prescriptions regarding structure and function of village organizations, and formulation of rules and regulations may overwhelm isolated villagers who have little familiarity with modern governance and bureaucratic procedures.

As in other countries, Cambodia's CF legislation requires the adoption of a "national" model of CF that presently exists largely as a concept, while actual CF groups take a variety of forms, and often reflect more traditional modes of communal organizations and resource use practices. There are an increasing number of local NGOs that are gaining experience in CF, fostered by donor small grant programs, and they may be in a position to facilitate interactions between Forest Administration territorial staff and forest-dependent communities as the national CF program unfolds. The successful implementation of CF laws and the new national CF policy will likely depend on flexibility on the side of both the government and communities, to find common ground where viable management partnerships can be established. There is also an urgent need for a broad-based effort to build the capacity of forestry staff, NGOs, local government, and community leaders to work together and restore and develop the nation's forest resources. The Japanese government has made a long-term commitment to training forestry staff, and is increasingly emphasizing community forestry components in curricula.

There appears to be growing concern in many rural Cambodian villages over rampant illegal logging, land encroachment, and commercial plantation expansion. This concern is reflected in local leaders, monks, and NGOs who are attempting to organize forest protection initiatives. Government agencies, NGOs, and local government units need to find ways to assist communities to organize forest protection activities, empowering them through the Land Law, Forest Law, and CF sub-decree. While Cambodia's transition to CF as a national strategy is still in an early stage of development, many important elements have been put in place over the past decade. If government and donor support can be mustered, several million hectares of forest land could be brought formally under community stewardship over the next ten years. This would likely require a long-term commitment from the donor community to finance the transition in the forestry sector, as the country retires logged-out concessions and transfers authority responsibilities to rural villages.

CHAPTER 8

Indonesia

Since the end of the Soeharto Era (New Order Government) in 1998, new laws and policies have emerged in Indonesia that support decentralization of governance and the rights of local communities over resources. For the most part, they have been promulgated at the central government level as part of new decentralization policies, however, district-level resource management policies are also being crafted. A few examples are provided below, but first it is helpful to understand the legal and policy history upon which they are being built.

After Soekarno and Hatta declared the Republic of Indonesia independent in 1945, the nation's new land and forest tenure laws and policy drew heavily on earlier Dutch colonial land tenure laws that recognized three ownership systems, 1) unwritten *adat* (customary or traditional) law, 2) land registered according to the Indonesian civil code, and 3) state land, based on the agrarian law of 1870 encompassing all land where no proof of ownership existed.[88] The independence struggle had destabilized government control of the forests, especially on Java where an estimated 500,000 hectares or 17 percent of Java's state forest land had been occupied by "squatters" or deforested. To help reassert state control, the new nation's leaders released six German and Australian foresters interned during the revolution to aid the Indonesian Forest Service in reasserting its control.[89] During the 1960s, the Basic Agrarian Law (1960) and the Basic Forestry Law (1967) the government continued to re-establish state authority, extending power to the Minister of Forests to designate land, determine use, and regulate management. The BFL granted only minimal recognition of the ancestral land rights of communities, and only if community use did not conflict with the mandate given to the Minister.

In the 1970s, a series of implementing laws were passed that legislated privileges for commercial timber extractors over local users. The Law on Forest Planning (PP 33/1970) precluded a role for communities in forest boundary setting and did not guarantee community compensation for land lost to concessionaires, essentially marginalizing policies and laws that gave some recognition to traditional management of ancestral forest lands.[90] As in most other Southeast Asian countries, in Indonesia throughout the 1970s and 1980s, indigenous forest communities were stigmatized as illegal practitioners of "slash and burn" agriculture and primary agents of deforestation. In part, this orientation was adopted to draw attention away from the growing forest loss caused by the commercial timber industry. As recently as 1993, a joint Decree from the Ministries of Agriculture, Home Affairs, and Transmigration (480/Kprts-II/1993) describes "forest dwellers that practice shifting cultivation as destroyers of the forest resource."[91]

Although Ministry of Forestry (MoF) policies emphasized centralized control and timber production throughout the 1980s and early 1990s, some effort was given to exploring community forestry as a management option, with the encouragement of development agencies. "In the early 1980s, the MoF and the Ford Foundation began to discuss social forestry programs, although it was not until 1985 that actual programs were launched, both in Java and the Outer Islands. The field experiences gained from these projects later informed the MoF as it formulated a nationwide social forestry policy."[92]

During the mid-1990s, the MoF passed a series of policies related to community forestry including the Forest Village Community Development program (SK691-PMDH) that obliged timber concessions to address some of the negative impact of their operations on local villages. These policies were based on paternalistic assumptions that communities needed the assistance of private logging companies who, in fact, lacked the motivation and skills to assist them. In 1995, a Ministerial Decree was made on Community Forestry (*Hutan kemasyarakatan)* granting limited user rights to rural villages in state production and protection forests, with an objective of regenerating degraded forest lands (Table 7).

The change in forest laws reflects an opening of civil society discussions regarding the topic. Pressure on the central government for forest sector reform grew as Indonesian NGOs developed their communication networks, research programs, and lobbying efforts through the 1980s. The Indonesian Forum on the Environment (*Wahana Linkgungan Hidup Indonesia-*

WALHI) and the NGO Network for Forest Conservation in Indonesia (*Sekretariat Kerjasama Pelestarian Hutan Indonesia – SKEPHI)* played a critical role in critiquing the patron-client relationships and advocating for local community rights, ultimately helping to place community forestry policy issues prominently on the national agenda.

It was only after the collapse of the Soeharto's New Order Government, that political pressures began to reshape governance and forestry policies. Under President Habibie's government, many statutory laws (*Undang-undang-UU)* and government regulations (*Peraturan Pemerintah-PP)* were passed to facilitate the socio-political transition and break-up old client-patron relationships.[93] Muslimin Nasution, the new minister for the Ministry of Forestry and Estate Crops (MoFEC) established a committee drawn from within and outside government to discuss forest sector policy reform. A series of workshops and meetings were held bringing Ministry officials together with NGOs, university researchers, development agency staff, local government representatives, and private sector leaders. The Communications Forum on Community Forestry (*Forum Komunikasi Kehutanan Masyarakat-FKKM*), a coalition of non-government representatives emerged to develop a draft forestry law for the MoFEC. They were disappointed, however, to learn that the Minister had also formed an internal drafting committee that produced Forestry Law UU41/1999, a document that directly acknowledges community-based forest management systems, while ultimately upholding the authority of the central government.[94]

FKKM had desired to vest full forest management control and ownership rights with the community. According to one analyst, "For policy makers who believed in state unity and centralized power and who believed that foresters' *raison d'etre* was to act as managers and guardians for the state, the idea of relinquishing total territorial control was far from persuasive."[95] Under the new law, communities are given equal access to forest land with big business and state enterprise. Indigenous peoples and ancestral domain claims are recognized under the new law. An important difference between Forestry Law UU41/1999 and the law drafted by the FKKM group was that MoFEC retained the ultimate control to oversee all forest territorial functions and tenure arrangements.

In summarizing the community forestry policy debate in Indonesia since 1998, Lindayati contends that the Forestry Law has extended important new rights over forests to communities, but more importantly has opened forest sector policies debate to the public. "The transformation of political systems and norms pervaded the forestry sector, providing more legitimacy to alternative social forest policy ideas that promote empowerment of local people. Debates over the 'appropriate' state management role no longer took place exclusively within the state domain but spilled over into the public sphere involving a much wider audience, including the media."[96] Table 7 briefly describes a few key laws and policies passed in the past decade affecting community forest management in Indonesia. They indicate a trend towards greater recognition of community resource rights, while retaining ultimate authority for the central government.

Despite the emergence of new laws that provide greater space for communities to play a role in state forest management, the new legislation also possesses a number of limitations. The Ministerial Decrees of the early 1990s designating Forest Villages essentially provided few resource rights to forest-dependent communities, but only made logging concessionaires responsible for providing development assistance. Since timber companies have little experience or incentive to implement rural assistance programs, most of these efforts had little lasting benefit on the ground.

The Ministerial Decree on Community Forestry provided villages opportunities to manage small timber concessions, but required them to form cooperatives first. Aside from a lack of familiarity with cooperative structures and operating systems, most Indonesian communities lacked experience in managing logging operations. The original idea of the decree was that small timber concessions will not be treated in the same way as in existing commercial timber concession areas. Rather, they should be managed through harnessing traditional knowledge and wisdom. Most communities who were awarded small timber concessions however, had been persuaded by logging contractors to maximize timber harvest.

As in other Asian nations, some actors in the private sector viewed community forestry as a way to continue commercial logging operations. The fundamental limitation in empowering communities with meaningful forest management rights through legislation is the reluctance of the MoF's unwillingness to divest itself of significant control. In a recent review of forest policy change in Indonesia, two researchers from the International Centre for Agroforestry Research (ICRAF) conclude:

> The failure of ministry policies and management practices to protect natural forest areas and improve the welfare of forest-dependent people is self-evident. There is increasing consensus among forest policy analysts in Indonesia that what drives the ministry is the desire to maintain control over as much of the Indonesia land base as possible.[97]

According to Fay and Sirait, much of the legal framework to devolve state forest management to communities and local government is already in place, only the political will to do so is lacking. "The mechanisms for such a fundamental shift in orientation exist, even with the current legal framework but the political will and leadership in forestry appear to be lacking. Clearly, an important first step is to redefine the state forest zone."[98]

The struggle over state forest lands in Indonesia at the national policy level involves the MoF and its support from powerful political and private sector actors, and the forces lobbying for greater political autonomy through decentralization. Consequently, the decentralization lobby has been one of the most effective forces in transferring forest management to local government.

Law on Regional Governance

While decentralization policies were welcomed after the fall of the New Order government and viewed as a way to facilitate local government and community-based resource management, the experience of the past five years has been disappointing for CF practitioners and planners in Indonesia. Decentralization has also not been uniformly accepted by senior Indonesian policy makers and political leaders. Like their counterparts in other Asian forestry agencies, many foresters believe that their departments should retain control over management decision making and are reluctant to devolve authority to local governments or even to provincial forest departments. In recent years, some of the initial policies that decentralized some forest management responsibilities have been rescinded, and the struggle for control between center, province, and community continues.

Decentralization policies have been widely criticized for accelerating deforestation as local governments attempt to capture a larger share of timber revenues, yet at the same time they have provided district councils and their community representatives with new authority over forest management, thereby bringing accountability closer to the local people. It is apparent that these national policies have been instrumental in creating the legal space required for local governments, particularly at the district level, to claim authority for state forest land management decision making.

TABLE 7: Recent Indonesian Laws Supporting Community Forest Management

Policy or Law	Strengths	Limitations
Ministerial Decree on Village Guidance and Forest Village Community Development (SK691/1991, SK610/1993, SK690/1995)	Made logging concessionaires responsible for providing development assistance to forest dependent communities	Gave no forest management rights to communities
Ministerial Decree on Guidelines for Community Forestry (SK622/1995)	Grants user rights to rural villages in state production and protection forests	Provides limited rights on degraded forest lands
Ministerial Decree on Community Forestry (Hutan kemasyarakatan) (SK677/1998, KMK31KptsII/2001)	Forest concessions can be given to people's cooperatives	New cooperative organizations must be established by the community to participate
Forestry Law UU41/1999	Acknowledges community-based forest management systems	Upholds the authority of the central government
Law on Regional Governance UU22/1999	Devolves greater authority to District/ Regency government for state forest lands management and regional autonomy, effective January 2001	Article 7 states that ultimate natural resource utilization authority is retained by the central government
Law on Fiscal Balance between Central Government and the Regions UU25/1999	Provides 64% of forest utilization rights levy to District/Regency government and increases revenue from forest royalties from 15% to 32%	Access to a larger share of timber revenues increases incentives for district government to accelerate exploitation of forests, with costs for communities.

Decentralization and Devolution of Forest Management in Java

In Indonesia, especially on the island of Java, where provincial governments have gained greater confidence and authority since the decline of the New Order government in 1998, local government is actively challenging the authority of the central government's State Forest Corporation (Perum Perhutani). Over the past five years, the Provincial government of West Java has issued decrees declaring that protected forest areas are under its sole jurisdiction. This action is in support of the local provincial government goal of placing 45 percent of all forest land under protection. Perhutani has retained control over forests zoned for production in West Java, however, the provincial policy also has established guidelines for Perhutani regarding felling, and requiring reforestation on barren lands with the involvement of local communities in management.

At the field level, however, Perhutani has frequently not complied with the provincial policy. Perhutani has generally continued its felling activities. According to local NGOs, Perhutani's community forestry efforts have largely failed to address community anger against the agency. In Desa Cibinong, outside of Bogor, communities have encouraged NGOs to pressure Perhutani to give them more rights and responsibilities for forest management under decentralization policies. Local governments are issuing orders requiring forest-dependent communities bordering lands to take responsibility for management. In most cases, Perhutani has not acknowledged these local policies and Perhutani retains strong political support from the center.[99]

Java's experience indicates the tensions that can emerge as communities and local governments, empowered by decentralization and CF policies, attempt to gain control over public forest lands. Efforts to secure local control over forests are often resisted by forest agencies and state corporations that had previously maintained unilateral control over the resource. Without clear leadership from central government policy makers and political will to support devolution transitions, conflicts will emerge that will continue to threaten the sustainable management of the forests, both on Java and in the Outer Islands. Perhaps, nowhere has the significance of forest decentralization policies been demonstrated as in Wonosobo District in Central Java where the District Council formulated a Community-Based Forest Resource Management.

The District of Wonosobo's regulation on CBFRM established its legal basis by referring to National Regulation Number UU 22/1999 on District Governance and on the new Forestry Law UU41/1999, as well as UU 25/1999.[100] The intent of the policy was to establish community management systems on all state forest lands in the district, authorized by the District head (Regent or *Bupati*). The CBFRM policy was designed to lay down the parameters for a transition from management by the State Forest Corporation (Perum Perhutani), a Jakarta-based parastatal timber management company, to community management. The policy allowed for the Regent to issue "Community-Based Forest Resource Management Permits" to local community groups authorizing them to continuously manage state forest lands. The policy also formally created the Wonosobo District Forest Forum, a multi-stakeholder coordinating mechanism to review the allocation of lands to community groups and monitor management activities.

CBFRM is based on several principles including:

- Sustaining Forest Function
- Sustaining Community Welfare
- Ensuring Social Equity
- Securing Community Resource Rights by Law

A significant aspect of the new policy is its claim to all state forest lands in the District (Chapter III, Article 6), placing them under the authority of the Council of District Representatives. In implementing the CBFRM policy the following five steps are undertaken including 1) site establishment, 2) community preparation, 3) authorization, 4) management, and 5) control. The CBFRM site selection process is executed by the District government, inventorying and redistributing all state forest lands through discussions with local communities, facilitated by the multi-stakeholder Wonosobo Forest Forum. Community preparation requires binding internal regulations be established for decision-making, conflict resolution, and other guidelines for management.

To apply for a CBFRM permit, the community group must submit an application including a site map, territorial specifications, group membership information, approved group regulations, and forest management plan. After passing a technical review, a CBFRM permit is issued to the community group by the district government, including a listing of the rights and obligations of the community and the district government. An objection period of three months is allowed, after which the agreement comes into effect. The CBFRM permit is valid for a period of up to 30 years, with an initial trial period of 6 years, and is extendable. The CBFRM permit cannot be leased, transferred or given as a loan to another group. A group member's rights are automatically transferred to his or her beneficiaries until the permit expires. CBFRM is not an authentication of land and forest ownership (Chapter V, Article 18).

The policy proceeds to enumerate procedures delineating protection and utilization zones, including setting aside riparian buffer areas, as well as allowable activities within each zone. In its official explanation of the CBFRM Policy dated December 15, 2001, the Secretary of Wonosobo District notes that:

> *Bearing in mind the success story of community forest management by the local community, Wonosobo District gives evidence that community is prepared and able to be directly involved in forest resource management in state forest areas... Therefore, timber-oriented forest resource management practices that have been serving as a state source of income, should be altered into community-based forest resource potential-oriented management...Bearing in mind that on the subject of realization of District Autonomy, particularly in forest resource management sector, it is time for the District of Wonosobo to administer its privilege...*

What is significant in this statement is that the District Council draws on the new decentralization policies to declare its control over the state forest lands. There are no references anywhere in the policy to the State Forest Corporation (Perum Perhutani), either in terms of its past role as state forest managers or in any future role, technical or otherwise. Following Indonesian law, the District policy was submitted to the Ministry of Interior for a three-month period of comment, after which it would be effective. After a period of consideration, the Ministry of Interior rejected the Wonosobo forest policy. It was apparent that the State Forest Corporation, seeing its authority for Java's state forest lands threatened, lobbied the Ministry of Forests and Estate Crops (MoFEC) to contest the policy with the Ministry of Interior. The Ministry of Finance may also have expressed concern over the potential loss of state revenues by devolving management authority to the district and ultimately to community groups. The Wonosobo District Council is continuing to lobby for further review of their policy. On the ground, a number of community forest groups in Wonosobo are already managing state forest lands under the terms of the proposed policy, on a *de facto* basis (Box 2). The State Forest Corporation has subsequently taken action to reassert its authority over forest lands being claimed by the District Government. By mid-2005, much of Wonosobo's forest lands and watershed continues to be a hotly contested domain.

National CF Program Implementation

Indonesia possesses approximately 60 percent of the forest land in the Southeast Asia region and an even larger percentage of the remaining primary rainforest.[101] With the majority of the land area classified as state forest "domain," issues of who controls this resource are of major national policy significance. Until the fall of President Soeharto's New Order Government in 1997, a national policy endorsing forest concessions was firmly in place. During the Soeharto era, 45 percent of the total forest area of 136 million hectares was zoned for timber production and leased to some 500 companies, with an average size of 100,000 hectares and some as large as 1 million hectares or more. The logging era proved to be disastrous for the nation's forests, despite 30 years of effort to design systems that could ensure sustainable utilization. Towards the end of the 20th century, forest loss was estimated at 700,000 to 1,200,000 hectares per year.[102]

In 1998, Indonesia experienced a number of changes that shifted the direction of forest policies and field level management. Indonesia, along with other Asian "tiger economies," entered a severe economic recession and the national currency, the *rupiah*, lost more than 80 percent of its value against the dollar. President Soeharto came under serious attack and was forced to resign amidst charges of corruption, including special privileges to friends operating in the timber sector.[103] At the same time, El Niño created a climatic pattern coinciding with forest fires that devastated much of Kalimantan and Sumatra. With the transfer of presidential power to B.J. Habibie, Abdurrahman Wahid, and then to Megawati Soekarnoputri, the political system and national policies changed rapidly including those guiding the forest sector. Supporters of community forestry hoped that the change in the political environment would create opportunities to enact CF policies and programs. However, what transpired was more complex.

In 2000, the Ministry of Forestry attempted to cancel all concessions and transfer them to state forest companies, while local authorities and pro-autonomy officials lobbied for greater decentralization. In 2001, Bob Hassan, President of the Indonesian Wood Panel Association (APKINDO) and a major timber concession owner was sentenced to two years in jail and many of Soeharto's cronies lost control over their concession areas. While many major timber concessions were closed, deforestation has continued and even increased in some regions as illegal logging became rampant and local government began leasing smaller logging areas. To the disappointment of many advocates, community forestry has not emerged as a major sector strategy in Indonesia, since the demise of the Soeharto government.

Official Coverage of CF

According to Government of Indonesia statistics, there are 591,000 hectares under community forest development. The Ministry of Forestry typically refers to this area as *hutan rakyat* (people's forests) occurring on privately titled lands that are held in the name of household heads. Village families practice a wide-range of horticultural and silvicultural management on these lands and the areas are accounted for statistically under the agricultural sector as they are often planted mainly with agroforestry species.

While Indonesian agroforests represent a diverse set of complex resource systems, community forest areas represented in Government of Indonesia statistics largely exclude vast areas of forest in the Outer Islands that have historically been managed by indigenous communities. This includes territories that have been under long rotation swidden cultivation, resin oil collection forests, hunting and gathering forests, and a wide range of forests that have been protected for religious and hydrological purposes.

Other recent official community forestry programs have been oriented towards small-scale timber harvesting (HPHH). As in the Philippines, many small communities have limited capacity to engage in commercial logging on a sustainable basis at the present time. Field reports from Kalimantan indicate that licenses for such small-scale operations have often been sold to timber companies.[104] As a recent CIFOR study reports there is a close two-way relationship between the local elites (regency government) and the forest entrepreneurs, and community involvement largely involves individuals who are powerful within the forest utilization network. The study concludes that: "The general tone surrounding the process of decentralization in Indonesia is one of overwhelming interest in the economic profits that can be gained from control over the utilization of forest resources in the regions."[105]

While the Outer Islands are sparsely populated compared to densely settled Java, indigenous communities are scattered in small settlements throughout most of the archipelago, even in very remote locations. A conservative estimate would suggest that at least half of the 135 million hectares of official forest land is utilized and under informal community management, suggesting that less than one percent of the nation's forests are formally recognized as CF territory by the government. It is helpful, therefore, to consider informal community management in Indonesia, as it continues to play an important role in resource stewardship in many areas, especially in fostering forest conservation and sustainable use.

Informal Coverage of CF

With over 17,000 islands covering almost 2 million square kilometers, Indonesia is one of the largest, most complex island nations in the world. While 60 percent of the country's 220 million people live on the densely populated islands of Java and Bali, representing 7 percent of the landmass, hundreds of ethno-linguistic groups are scattered across the islands and forests of the archipelago. Much of the population of the Outer Islands are dependent on natural resources for their survival, relying on a wide array of agricultural, horticultural, agroforestry, and forest management systems. For centuries, resources have been managed through indigenous cultural institutions with unique sets of customary law (*adat*) and tenurial arrangements.

While no statistical data is available, it is reasonable to assume that much of the area identified as "official" state forest land in the Outer Islands falls within the ancestral domain of some cultural community. The Ministry of Forestry has little capacity to manage areas claimed by the state. With 50,000 personnel of whom only one-third are full-time staff, and with half of the professional foresters based in Jakarta it is unrealistic to assume that MOF has a presence on the ground to manage the forests. This works out to perhaps one forester for every 100,000 to 300,000 hectares.[106] By contrast, there are an estimated 30 to 40 million people living within or near forest lands who are dependent on them in varying degrees for their livelihood. The communities represent 250 indigenous cultures, speaking over 600 languages.[107] It has been these communities that have managed the forests according to their unique traditional practices for hundreds of years, and in many areas they continue to do so.[108]

Since land titling has not yet been done on a large scale in many parts of the Outer Islands, large tracts of indigenous land have been leased to concessionaires. Nonetheless, in many areas communities continue to manage these lands, as well as areas where no concessions exist. This is particularly true in Papua. While indigenous resource rights and community management are recognized in the 1870 agrarian law and the 1945 constitution, and more recent policy dialogues are focusing on communal rights, for most of the past century, national "development" priorities, including commercial logging, have been given precedent.

A study by the Communications Forum for Community Forestry (FKKM) identified some 18 distinctive types of indigenous community forestry systems. These vary widely in terms of organization, production and conservation goals, as well as the ecological contexts in which they operate. They vary from upland swidden and forest fallow cycle management, to intensive agroforests, coastal mangroves, sago palm groves, to forest gardens, and natural stands. While they are indigenous forms of management, approaches can be dynamic with ongoing efforts to integrate new production techniques to enhance income flows and output. Many of these indigenous forest management systems operate on forest land that is claimed by the state.

The process of state recognition of indigenous forest management systems and their ancestral claims of tenure has not taken place on a broad scale in Indonesia, despite language in the Basic Agrarian Law, Constitution, Forest Law, and more recent community forestry regulations. An exception to this general situation is the case of Krui in southern Sumatra. Local communities have managed approximately 54,000 hectares of forest gardens (*kebun damar*) with a resin-producing tree (*Shorea javanica*) for over a century. Logging companies and rapidly expanding oil palm estates in the mid-1990s began posing a growing threat to this indigenous forest management system. Since the forest gardens were within the state forest domain, and the communities had little formal recognition of their communal rights to the forests, they felt their control over the resources was in jeopardy. With assistance from FKKM and the International Centre for Research in Agroforestry (ICRAF), the area was declared a "special use zone" (KdTI) in 1998. The new community forestry designation grants Krui communities control of ancestral *damar* forest reserves under customary *adat* institutions and laws. Private households have long-term rights to utilize their forest patches, but they cannot be sold outside the community. There is an urgent need in Indonesia to extend formal recognition of the forest management rights of local communities. The challenge is establishing the political will to do so, while creating capacity to implement such a policy throughout an immense and complex nation.

With the demise of the New Order government in 1998, there was hope that community forestry policies and programs would emerge under more democratic administrations. Some analysts felt that this would create an enabling environment where formal management rights for Indonesia's vast forest areas could be transferred to local communities throughout the country. In retrospect, the transition to community forest management as an important national approach to forest stewardship has not taken place. The Wonosobo case study illustrates the reluctance of state corporations to share meaningful control, and their political ability to block devolution, even where local government supports community management (Box 2). Decentralization policies enacted in early 2001 have generally led to a vesting of timber exploitation rights to regency (district) level political leaders, not to communities, and logging has actually accelerated in many areas as a result, especially in the Outer Island.

While commercial timber harvesting reached its peak in the 1970s and 1980s, over the past ten to fifteen years it has been the clearing of forests for large estate crops that has dominated landscape change in the Outer Islands. The government has largely failed to control the burning of millions of hectares of logged over land by entrepreneurs interested in establishing oil palm and other commercial crop estates in Sumatra and Kalimantan. Further, general skepticism of community forestry as a viable national forest management has been retained by many staff within the Ministry of Forestry in much the same way it was perceived during the New Order Government. Yet, on the ground communities in many areas continue to be actively involved in forest use and protection. Until Indonesia is able to address the disjuncture between its policy emphasis on forest utilization and revenue generation by private and state enterprise, and the tens of millions of rural people who retain ancestral domain claims and seek to manage it for multiple purposes, the nation's forestry sector will remain chaotic, with considerable conflict between policies and practice.

While government support, especially from the Ministry of Forestry, has been limited, community forestry has received a growing boost from civil society at large. The emergence of networks like FKKM has provided national platforms through which community forestry issues can be discussed. At the provincial and regional level, the strengthening of cultural groups has provided new political channels through which resource issues can be addressed. In Indonesian Borneo (Kalimantan), for example, the growing influence of Dayak associations has helped guide new social and ecological movements in that region.[109] Local NGOs emerging from within the cultural communities are engaged in a growing number of activities including community mapping of lands, facilitation of dialogue with local government, and working as advocates for communities with forest land conflicts. Other major cultural communities including the Batak, Minangkabau, Bugis, Torajans, and Papuans are organizing with growing effectiveness in efforts to address pressures from outside interests seeking to exploit local natural resources. If recent trends continue, it would appear that the best hope for community forestry will come through programs initiated by the country's cultural communities with the support of local government, NGOs, and local universities.

Across Java, communities are expanding indigenous agroforestry patches and forest gardens in an effort to improve the productivity of their land resources. These systems are based on traditional mixtures of species, and form multi-storied forestry ecosystems that include herbs, medicinals, climbers, fruit trees, and timber trees. In some villages nearly two hundred species of plants are present. The forest gardens, referred to as *wono dusun* (village forest in Javanese) and *talon* (in Sundanese), are extremely profitable, often generating more income per hectare than irrigated rice fields, while providing greater protection against disease, pest, and market risks due to their diversity. While there has been no attempt to distinguish changes in forest cover under community and state management, research indicates that forest cover in Java increased by almost 600,000 hectares (from 1.27 million hectares in 1985 to 1.87 million hectares in 1997), largely due to the expansion of community-based agroforestry and forest gardens.[110] This process is continuing as communities in some parts of Java are taking over degraded state forest lands and establishing mix-forest gardens.[111] It should be noted that the increase in these private, community forests on the island of Java comes with no investment from the state or any development agency.

The reform of the State Forest Corporation (Perum Perhutani) and the devolution of forest management to communities could allow for the transformation of several million additional hectares of degraded forest land on Java if an enabling policy framework were established. This transformation in the forestry sector in both the inner and outer islands of Indonesia is being resisted by a number of stakeholders including many within the forestry agencies, as well as private sector interests involved with the timber industry, plantation development, mining, and other extractive enterprises. In 2004, Perum Perhutani, with the support of the central government, issued an order for the community-based agroforestry gardens in Wonosobo to be cleared and in April 2005 the Ministry of Interior nullified the Wonosobo District Regulation on Community-Based Forest Management. Such actions reflect the intention and ability of the state forestry establishment to frustrate community initiatives and exert their authority over the nation's forest resources.

While the legal and policy orientation of the Government of Indonesia to its public forest lands can not be generalized from a single case like Wonosobo, other experiences from that country indicate that CF has not gained much momentum as a formally accepted management strategy. While some pilot activities are being undertaken and research continues, there is little indication that a broad-based devolution of state forest land management authority to rural communities in Indonesia will be launched in the foreseeable future. While progressive policies and laws may be approved, and donor-sponsored CF programs initiated, without the political will to reform the sector, the old paradigm will prevail bringing with it inequitable and unsustainable use.

BOX 2: Communities Transforming Forestlands in Java, Indonesia

This case study illustrates the ways in which some upland communities in Java are taking over management of degraded forest lands with the support of their local governments. While the impact of reforestation activities undertaken by communities have successfully transformed poorly stocked and eroding pine forests with dense, multi-storied agroforests, the take-over by communities and the district council has encountered strong opposition from the State Forest Corporation that has managed the land for decades. This process represents some of the opportunities and challenges Indonesia faces as it has attempted to devolve forest management responsibilities to communities and local government agencies since the fall of the New Order Government in 1998.

Intensive Land Use. Wono dusun is a common resource unit in Java that has multiple uses—agriculture, forestry and livestock raising. This intensive agroforestry system provides local communities greater economic stability while maintaining ecological sustainability.

The Javanese in Wonosobo District of Central Java have a long established tradition of agroforestry—*wono dusun* in the local language. Much of the private lands are managed under a variety of forestry and agroforestry systems that are often characterized by a remarkably high level of species diversity. Research on traditional *kebun* (agroforest) and *talun* systems (forest garden) in West Java have shown farmers may possess several hundred species of trees, climbers, shrubs, and herbs within a single village. These systems are highly productive and generate a wide range of products including medicines, fibers, timber, food, ornamentals, fodder, fuel and other useful goods. While multi-storied production forests have been a common feature of the Javanese landscape for centuries, during the Dutch colonial period millions of hectares of coastal, lowland, and upland forests were placed under a state forest corporation (SFC) currently referred to as Perum Perhutani. Perhutani has a

FIGURE 3. Process for Approving and Monitoring Community-Based Forest Management Permits, Wonosobo

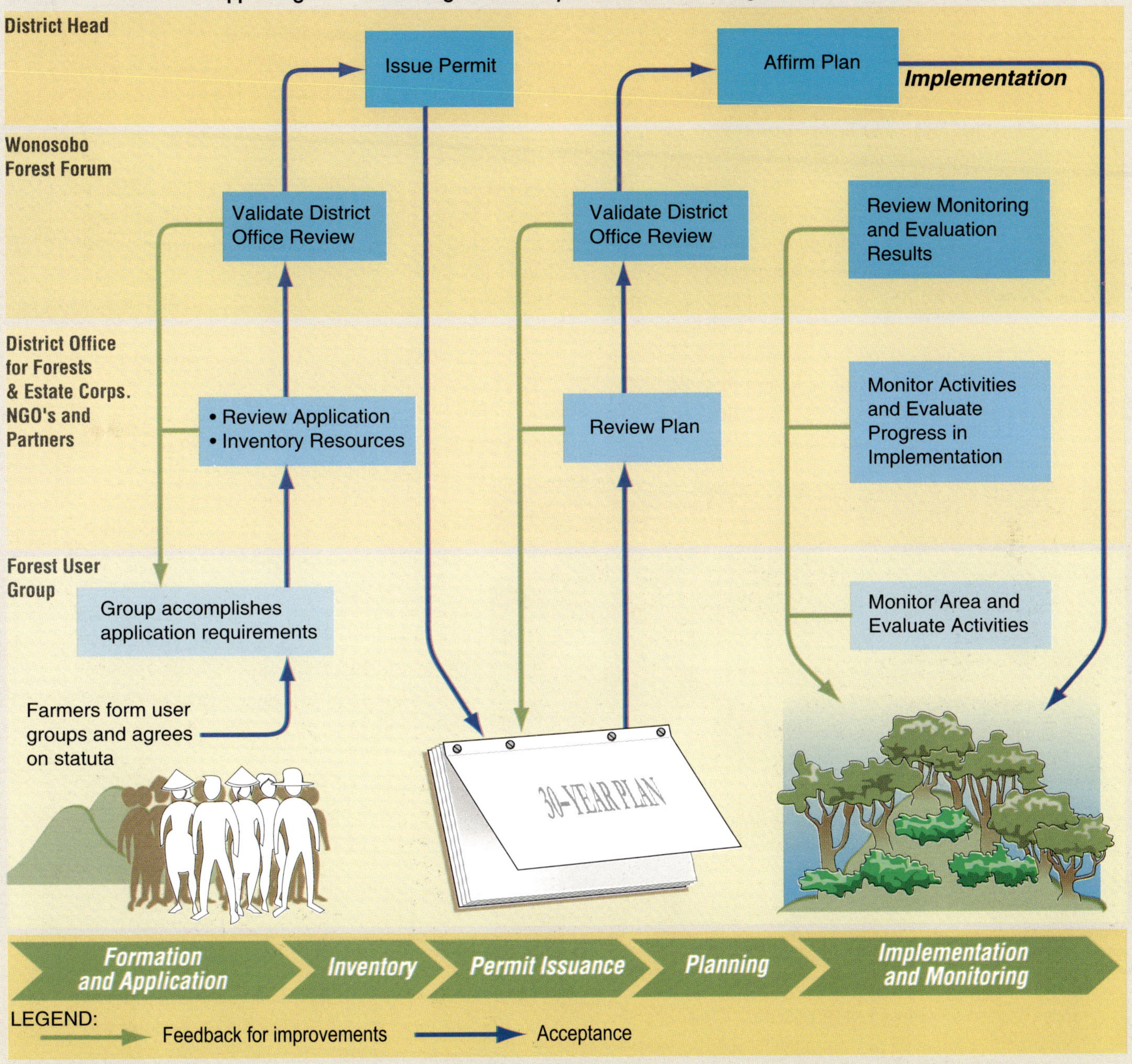

long tradition of allowing communities to plant food crops in newly planted forest areas. Under this system, known locally as *tumpang sari* (or more generally as *taung ya*), as the canopy closes after three years, local farmers must cease cultivating the areas. With a population of approximately 100 million, land pressures on Java are intense, and tension between Perhutani and rural communities runs high. Farmers on Java increasingly seek long term rights to use and co-manage forest areas.

This project was an attempt by a local NGO (ARuPA,) that is primarily comprised of young students from universities in Yogyakarta, to work with communities, local government, and Perhutani to develop a collaborative forest management program. The project is ongoing in Wonosobo District, situated midway up the volcanic slopes in south central Java. The original pilot began in Bogoran village in an effort to test the implementation of new mechanisms crafted under the CBFM District Regulation (Figure 3). During the planning phase, local farmers admitted that they were already establishing agroforests on neglected state forest lands in the villages. While they knew this had been illegal in the past, they hoped that their planning efforts to improve the management and productivity of the degraded pine forests could be formally recognized and approved under the new district regulations and that some equitable sharing agreement could be developed with government.

In 1999, soon after the fall of the New Order government and with the uncertain political environment, a number of farmers began felling degraded pine forests, reportedly with the encouragement of local Perhutani staff. While Perhutani requested that the farmers plant mahogany, and said they would allow for cultivation for three years, they did not provide any seed or planting materials. According to local villagers, "We did not have a clear contract with the Perhutani's district office, so we are not following what Perhutani prescribes. We are now planting whatever we want, but we are looking after the trees that we have already planted." The villagers hope that eventually an agreement will be worked out with Perhutani so that their formal rights and responsibilities are clarified. As one forest farmer noted: "Never mind if there is no exact sharing percentage as long as people get a bigger share. After all, we are the ones working on the land."

Women and upland resources. Women form a group to earn cash from helping tend the land of other villagers.

In the meantime, the Wonosobo District Government and its legislative branch (DPRD) issued a District Regulation on CFM (Perda 22) that provides the basis for giving greater tenurial rights to communities working on state forest land. Under this regulation, the District Chief (*Bupati*) will provide 30 year tenurial agreements with communities through the District Forest Office. A Wonosobo Forest Forum (WFF) was created to implement the District Regulation on CBFM and to monitor its application. The WFF is comprised of local government representatives, community leaders, NGOs, and university advisors.

Wonosobo District, with well-connected road networks and electricity in villages, has developed a vibrant wood processing system based on fast growing species that are even being exported to the Outer Islands for use in particle board manufacturing. The forest farmers of Wonosobo currently enjoy support from district government after the decentralization law. Many young adults have finished high school and some are going to college, yet are remaining in the area and are often active in organizing environmental initiatives and working with NGOs. The new generation is interested in exploring how they can get optimum benefits from the state land that their families have tilled as tenants under the *tumpang sari* program. They are frustrated by the difficulties in getting benefits and resource security given the long history of state ownership and management on Java. Yet, the Perum Perhutani has strong allies in the capital and has strongly opposed any concessions to the forest farmers. The Ministry of Forestry and Perum Perhutani are both contesting the District Government regulation as it directly threatens the historic role of Perhutani and the authority of the national ministry.

Central elements of Indonesia government have gone beyond what they initially backed through the decentralization process. Always sensitive to the Outer Islands and wary of the growing use of authority in provinces, the administration sees the need to tighten the controls on resource allocation. In Java, district action is affecting primarily semi-state interests and so is difficult to tolerate. Wonosobo is an example where there have been efforts to find a way to work together. Wonosobo presents a classic test case regarding the political will of the national government to decentralize forest management responsibilities to local governments. In early 2005, however, the CBFM District Regulation was revoked by the Ministry of Interior. ■

CHAPTER 9

Philippines

The Philippines is considered to be a center of experience with community forest management in the Asia region. This status can be attributed to early attempts to devolve state forest management rights to communities through a multiplicity of policy and programmatic approaches. Strong community stewardship traditions are also linked to insurgency movements of the past three decades that have denied state authority over natural resources and allowed communities to retain control over many upland watersheds. Before describing the forms community-based natural resource management has taken in the Philippines in the past thirty years, it is useful to review briefly the history of forest management in that country.

Historical Policy Context

Historically, the Philippines has been a nation with varied and rich forest ecosystems inhabited by over a hundred ethno-linguistic groups of upland and coastal communities.[112] In 1521, Magellan landed on the island of Luzon and claimed the entire Philippine archipelago for the Spanish Crown. Although this policy had little effect on communities during the early colonial period, it established a precedent for state control of land that persists to the present time.[113] According to Lynch, "the legal effect of Magellan's gesture was to convert all indigenous occupants of the still unexplored archipelago into squatters."[114] The Cooper Act, also known as the Philippine Bill of 1902, systematized this state control through the American codification of Spanish land laws and is now enshrined as the Regalian Doctrine. This policy strengthened the role of government in allocating resources, making the forestry industry fourth in terms of production and a major contributor to government revenues.[115]

The land and forest rights of the local population eroded further as Spanish colonial policies and laws were implemented. Unless individuals and communities documented their claims to land and forests, the land was considered to be the property of the Spanish sovereign. The clearing of the forests and their conversion to agriculture took place at an accelerated pace during the 17th and 18th centuries, initially in the lowland areas. The Philippines was one of the first nations in Southeast Asia to experience massive deforestation. Extensive tracts were cleared in the central Luzon plain, Cebu, Negros, and other Visayan Islands to provide fuel for sugar cane refining during the 19th century. The island of Cebu was so badly deforested that logging was banned as early as 1870. In reaction to rapid forest loss, the first Philippine forestry bureau was established in 1863. The institution was given authority to manage much of the nation's land area, a policy that continued to be followed by the American colonial administration. This power of designating lands as private or public and increasing ability to enforce decisions, under the Philippine Constitution, gave to the legislature – *de facto* its elite – the power to acquire further land as well as the influence to gain leasing rights over public lands. Though the Bureau of Forestry became the "sole owner" of forest lands, it only developed limited revenues which the elite controlled.

In 1900, George Ahern was appointed the first American Director of the Philippine Bureau of Forestry, with 20 million hectares of forest land. Despite the presence of a technical agency to oversee forest use, however, unsustainable logging was rampant. In 1905, for example, an American forester claimed that if there was a revolution in the Philippines it would be due to the failure of the Forestry Bureau.[116] While sustainable management of timber concessions was required under twenty-year renewable leases, the agency was unable to enforce compliance and by 1914 the Bureau reported many logging companies were virtually clear cutting the forests. The inability of a national forest agency to implement sustainable management policies was repeated throughout much of Southeast Asia over the next 100 years.

After independence was achieved in 1947, the new government of the Philippines viewed logging concessions as a key element in their economic development strategy, as did many countries in the region. Justified on the basis of generating revenues for national development, forest concessions were often used by political leaders, such as Ferdinand Marcos, to build alliances with other powerful families. During the

Marcos administration, over 400 timber license agreements (TLAs) were issued. Between 1980 and 1990, the Philippines had one of the highest deforestation rates in the world, losing 3.5 percent annually.[117] Even in the face of rapid deforestation, a national allowable cut of 15.5 million cubic meters was set during the peak logging period in the mid-1970s. The process of deforestation was often initiated by logging, followed by immigration on logging roads and the further conversion of upland watersheds into agricultural lands. Ironically, throughout the 1960s and 1970s, the government frequently blamed indigenous communities and migrants, referred to as *kainginero (*or practitioners of swidden cultivation), as the primary cause of deforestation.

Evolution of CF Laws, Policies, and Programs

In 1974, the Government established the first policy that provided some recognition of community tenure rights under the Forest Occupancy Management Program (Table 8). However, it was never widely extended. Additional programs began to be initiated throughout the 1980s, as the government gradually crafted new responses to tenure insecurity concerns and unsustainable natural resource management, especially in upland watersheds. Despite the move towards devolving some natural resource management rights to communities, by 1988, 1,245 forest concessions and grazing leaseholders still controlled 5.8 million hectares of public land representing one-third of the nation's public lands. While several thousand, usually absent, families controlled these vast resources, over 17 million people populated the upland and only a small proportion saw their situation improve through palliative projects. [118]

From 1960 to 1999, reforestation projects were undertaken covering 10 percent of the uplands, with a focus on key watersheds and logging concessions as part of the DENR's effort to improve forest management. Most reforestation projects implemented by the government were financed with bilateral and multilateral loans. Only for a short period in 1984 and 1993-1996 did the non-government organizations and the private sector account for more than 50 percent of the total area of 1.7 million hectares reforested. The DENR reports that less than 50 percent of the replanted area has been effectively managed.[119] The ADB estimates that the survival rate of trees in upland reforestation projects is 30 percent,[120] while others sources place long-term survival at 21 percent.[121]

TABLE 8: Timeline of Community-Based Natural Resource Management Development in the Philippines

1963	Revised Kaingin Law defined occupants as residents and cultivators (RA 3701)
1971	Kaingin (Swidden farming) Permits Issued for Forest Dwellers
1974	Forest Occupancy Management Program
1975	Revised Forestry Code of the Philippines (PD 705)
1979	Family Approach to Reforestation (FAR) and Communal Tree Farming (CTF) with renewable 25-year lease
1982	Integrated Social Forestry (ISF) Program initiated
1986	National Forestation Program (NFP) $240 million loan ADB and OECF
1987	President Cory Aquino guides New Constitution – Dept. of Environment and Natural Resources (DENR) Created
1989	Community Forestry Program initiated
1990	Ancestral Lands and Domains Recognized (under New Constitution - Article II Section 22)
1991	DENR Policy Devolves some NRM Functions to Local Government (RA 7160)
1992	DENR Issues Procedures to Demarcate Ancestral Lands (Administrative Order No. 2) Congress passes the National Integrated Protected Areas System Act
1995	Community-Based Forest Management (CBFM) as National Strategy for Sustainable Forestry and Social Justice (Executive Order No. 263).
1996	CBFM Implementation Procedures Approved (DAO 29)
1997	Certificates of Ancestral Domain Claim (CADC) cover 1 million hectares Congress passes the Indigenous Peoples' Rights Act
1998	Guidelines for CBFM within Mangrove (DAO 98-10)
1999	Formation of a Presidential Task Force on Ancestral Domains (MO 52)
2003	Over 5 million hectares of forest lands allocated through some form of community tenure
2004	Promoting Sustainable Forest Management in the Philippines (EO 318)

Many reforestation projects, though ostensibly "community-oriented," failed to involve communities effectively and local resentments ultimately led to efforts to sabotage reforestation projects through intentional burning. This continues today in some CBFM areas where communities have not been able to internalize the sense of rights that they have to the area. In other areas, where plantation species are used, this invites "illegal" cutting by individuals where trees have survived. In many areas reforestation programs were misguided in their implementation and lacked community focus; time was lost in giving clear rights and financial returns to communities.[122]

The greatest success in increasing wood production is where small farmers have planted the perimeter of their land or small parcels of land with *Gmelina arborea*, as an added contribution to the country's wood needs.[123] Though the government strategy is beginning to include such community participation, the damage to ecological services, particularly water availability and quality, is so far advanced that it will take decades of work by upland communities and government to restore the environment.

While community forestry initiatives were emerging very gradually under the Marcos administration during the early 1980s, there was little political will to invest heavily in reform programs and impact on the ground was very limited. With the EDSA Revolution and the emergence of the "People Power" movement in 1986, under the leadership of Cory Aquino, it was hoped that forest policy change would occur rapidly in the Philippines. In fact, President Aquino lacked experience with government and policy making and entrusted policy to her cabinet. Secretary Ernesto Maceda, who handled natural resources, did much to reinforce traditional logging leases during his six-month term. After Secretary Fulgencio Factoran, a human rights lawyer, took over, new policies began to emerge including a ban on log and timber exports, a moratorium on TLA issuance, and a shift to community-based forestry.[124] Despite the move to participatory resource management, administration of the public forest domain continued to be through TLA from 1960 to 1999.

The logging industry did not take this shift from "pro-logging" to "people's forestry" without a fight, and put considerable pressure on the DENR to cease its efforts to increase levies for extraction and utilization charged to concessionaries. The DENR responded by drawing in a huge amount of project financing from multilateral and bilateral development agencies for community forestry programs, totaling $856 million between 1988 and 1992. In an unusual tactic, the DENR requested donors to include conditionalities in the loan agreements that increased logging fees and banned felling in virgin forests. As Ross notes:

> *Almost none of these conditions were drafted by the funders, rather, they were requested by the DENR leadership and negotiated with funders in advance. Factoran and his allies realized that to change Philippine forest policies they would have to fight the logging industry and its proponents, particularly in Congress... The pro-reform faction in the DENR turned to internal funders for political help...partly through the use of conditionalities...the ban on logging in old-growth forests, which was tied to the USAID grant, was not simply designed to protect the forests but to undercut the industry's economic strength by denying loggers access to the country's most valuable timber...*[125]

By relying on donor conditionalities, the DENR had the additional benefit of locking in the reforms, as such policies could not be changed without the approval of the World Bank, or ADB, or USAID. Another major achievement during Secretary Factoran's tenure was the recognition of Ancestral Land and Domains that provided the basis for claims covering several million hectares in the Philippine uplands by 2003.

In 1992, President Fidel Ramos was elected, and Angel Alcala, then president of Silliman University, was made DENR Secretary. The forest policy initiatives of the Aquino Administration were largely continued under the new administration with community forestry as the "flagship" program. Yet, reports by those who closely monitored activities of the DENR during the period indicate that corruption had grown, political manipulation had increased, and that commitment to implement CF had waned.[126]

In 1995, Victor Ramos took over as Secretary for DENR. According to one report, "Overall, morale was low because of incidents of corruption within the department... Erring personnel were charged and suspended. Questionable orders were reviewed and the Industrial Forest Management Agreement (IFMA) program was suspended in 1995...one big change Ramos observed when he came back to DENR was the meddling of politicians in the appointment of DENR field personnel."[127] By 1995, the number of Timber License Agreements had declined to 29, from 230 in 1977.

In retrospect, while the change in the formal policy atmosphere at the DENR between the tenures of Secretary Factoran, Secretary Alcala, and Secretary Ramos was minimal, the alteration in informal orientation and work environment, especially as it impacted CF programs and strategies, was quite dramatic. According to Secretary Ramos: "Even if there are new laws in forestry, much has to do with implementation. This boils down to leadership and the values the leader espouses."[128]

Under Secretary Ramos, DENR policies stressed community-based forestry and proper pricing of forest resources, and controlling illegal logging. Strategies to fight illegal logging changed, however, from conventional policing tactics to a new focus on the Community Forestry Program (CFP). Ramos noted, "community forestry can cancel out illegal logging. Instead of talking about volume confiscated, we'll talk about communities and areas where community forestry can take root."[129] Another important change in the legal and policy environment was the new emphasis on decentralization. While the DENR had sought out the NGO community as partners in the late 1980s, the new focus was on working with local government. Ramos noted, "We're seriously courting LGUs because they're closer to the problem than us. They should be able to arbitrate and translate policies into ordinances."[130] While decentralization offered an opportunity to partner with local authorities and develop management rules and strategies based on site-specific knowledge, it also posed a risk of local political elite imposing their personal priorities above national goals.

With the election of President Estrada, the political will to implement CBNRM disappeared. Secretary Antonio Cerilles stopped the issuance of CADC and the resource use permits included in CBFM agreements were suspended for over a year. Overnight, communities' resource extraction activities became illegal. It was a period of very limited policy development and program implementation, demoralizing the newly community-oriented line staff. During this period, progress in promoting the devolution of resource management rights was largely made through the transfer of numerous responsibilities to local government. Only with the departure of Secretary Cerilles did the damage done to CBFM become rectifiable.

With the swearing in of President Gloria Macapagal-Arroyo new social commitments with policies and programs emerged. The result was a very strong political effort to give Certificate of Ancestral Domain Titles (CADT) to communities. Under the new administration, the Presidential Task Force on Ancestral Domain was formed. While vast areas of upland forests were formally transferred to cultural communities in the uplands, management capacities remained limited. The new administration also determined that a review of CBFM plans and programs was needed before any policy decisions were taken. For a short period of 18 months, Secretary Elisea Gozun again opened up the DENR to a level of transparency and engagement with civil society. The results of the CBFM reviews, however, were not as convincing to national economic planners as the prospects for further large-scale resource extraction.

CF Policy Development Process

CF policy and program development has been fostered by a variety of groups and multi-institutional mechanisms for over two decades. In the early 1980s, with support from the Ford Foundation, an Uplands Working Group was formed bringing together

Impact of reforestation. Many reforestation projects in the 1980s, though ostensibly "community-oriented", failed to involve communities effectively and local resentment ultimately led to efforts to sabotage reforestation projects through intentional burning.

individuals from the Bureau of Forest Development (predecessor to the DENR), the University of the Philippines, the Ateneo University, and De La Salle University. As the years progressed, a number of NGOs joined civil society dialogues regarding the involvement of communities in resource management.

In 1994, the Philippine Working Group (PWG) was formed to foster Community-Based Natural Resource Management (CBNRM) and was comprised of an informal group of professionals and practitioners. Over the past ten years, the PWG has worked in 60 sites around the country to monitor the implementation of national CF strategies, assess their strengths and limitations, and to facilitate a dialogue among national planners, local government, NGOs and communities. Activities have included site visits that dealt with specific resource management concerns, community mapping assistance, and training, as well as consultations with national and provincial planners regarding policy and program strategic needs.

On-going Policy Concerns

While CBFM has strong support from the current administration of President Gloria Macapagal-Arroyo, and a broad legal basis for implementation, the strategy confronts a number of new and ongoing challenges. An analysis by the PWG identified the following "shadows" in one of Asia's groundbreaking CF programs.[131]

CBFM continues to stress commercial timber extraction, despite urgent needs for watershed protection. Given the rapid loss of forest cover in the Philippines, and the important environmental services they provide to both local communities and downstream populations, expectations for revenues from extraction are considered misplaced by many followers of the CBFM program. Emphasizing commercial utilization, even if done by communities, continues to place unsustainable pressure on the diminished resource base, while creating opportunities for corruption and exploitation. Demand for CF policies that are extractive in orientation may be generated by ongoing interests of timber companies that seek to work through communities and their backers. They may also be encouraged by development agencies that seek to justify loans by projecting economic returns from timber.

CBFM program support is subject to political priorities of DENR leadership. As the history of CF cited in this review indicates, the political will to implement CF in the Philippines has varied sharply over the past two decades, reflecting the political orientation of DENR's leadership. The position of DENR Secretary appears to be most influential in determining the rate, quality, and transparency of CF implementation. Unfortunately, this creates an uneven pace for implementation, accelerating under some administrations, and slowing under others, sending conflicting messages to DENR field staff, local government officials, NGOs, and communities.

CBFM has limited funds for national program implementation. After being flooded with nearly $1 billion in development assistance financing for CBNRM in the 1990s, the DENR currently has little external support for CF. This reflects changing donor priorities, as well as dissatisfaction with DENR progress, especially during the administration of President Estrada. The funding crisis now confronting the Philippines' national CF program clearly illustrates the problems Asian government forestry agencies face in depending on donor financing for forest sector management transitions that require decades to implement. CBFM is still competing with other "national strategies" for sustainable forest management. This has diminished the attention it receives from policy makers. The National Integrated Protected Area System Act-NIPAS (1992) and the Watershed and Ecosystems Planning Framework (1999) were used as a protection against mining in many areas, without fully understanding the consequences for community management. The Sustainable Forest Management Act has been delayed for over 10 years and continues to get held up by Congress.

The implementation of CBFM policy is now dependent upon local governments who acknowledge that they have been given the responsibility but lack the finances and authority to control CBFM projects. CBFM is unable to quickly demonstrate the impact of the policy on improving forest cover in the Philippines. While proponents believe CF has helped stabilize many upland watersheds and forest areas, the process of capacity building requires long-term investments in institution building. Corresponding changes in improving vegetation and forest conditions also need substantial periods of time. Yet, politicians are putting continuing pressure to demonstrate that CF policies are working.

CBFM has faced problems demonstrating widespread poverty reduction desperately needed in the country. All too frequently, communities participating in the CBFM program do not possess copies of their agreement, do not develop work plans, nor review, improve and update them. The short-term nature of donor-supported project funds and the lack of an adequate market base for productive activities have frustrated the emergence of viable, forest-based livelihood activities. As a consequence, the 5 million hectares under community tenure have, for the most part, not been brought under productive management with corresponding impact on household economic conditions. The present economic policy is increasingly in conflict with CBFM strategy, as it seeks out large-scale investments that promise financial returns that can address the national deficit, rather than supporting the slower process of economic development from below.

As the Ancestral Domain policies and programs originated from the DENR as part of the broader CBFM strategy the application of CBFM and the Certification of Ancestral Domain Claims can be confusing for practitioners, and may create opportunities for manipulation by vested interests. The present requirements for annual plans need to be rationalized, simplified, and shared by both the CBFM and Ancestral Domain programs. There is also uneven coverage of and support for CBFM programs. In most areas, CBFM projects are driven by special donor agency support and non-government action. Many remote communities have never been visited by any government officials (other than the military) or provided with basic services, and have never discussed their forest and land tenure concerns and resource management problems with government representatives.[132]

Provincial planning is beginning to incorporate peoples' participation in resource management. Where such incorporation occurs, provincial planning generates direct social investment in resource management that otherwise starts with speculative private sector investment. This social investment becomes more self-reinforcing rather than self-limiting and catalyzes positive feedback. Local government increasingly recognizes the importance of cultural practices in managing upland areas. A more active response is needed to meet migrant land use practices, insecurity and unsustainability, particularly as migrants also dominate the local positions in government as *sitio tanod* (village official) and barangay councilors. Comprehensive land use planning needs to be operationalized with sufficient inter-provincial and regional discussion so that people and resources are not fragmented.

The radical shifts in resource management orientation under way in the Philippines over the past twenty years must struggle with centralized policies that have evolved over the past 400 years and cannot be quickly reversed. Given the recognition of the need to decentralize government, the increasing acknowledgement of cultural communities and their place in society, international concern over livelihood needs, and the loss of widespread ecological services, community management is increasingly accepted in the Philippines as a key to the restoration of the environment and productivity of resource in the coming century. The passing of strong laws supporting ancestral domain claims is an achievement of special significance. As one analyst notes, "the move to get ancestral domains recognized is part of a much more profound campaign for land and human rights than the relatively narrow forest-based tenure instruments that have been tried from time to time."[133] In fact, it is likely one of most significant changes in state orientation to land ownership in the Philippines since the Regalian doctrine was put forth over four hundred years ago. This is not to detract from the national Community Based Forest Management strategy, which has overlapping goals with the Ancestral Domain Claim program. Nonetheless, the two have distinctively different characteristic and need to be assessed and monitored separately.

National CF Program Implementation

The Philippines is depicted to be a Southeast Asia pioneer in the development of community forestry as a national strategy for state forest management. Community management agreements were first formally issued in the Philippines in 1982 through certificates of stewardship under the Integrated Social Forestry (ISF) program. The ISF program officially adopted social forestry as a forest management and development strategy in the uplands. This program also incorporated earlier efforts to involve upland communities in forest management in the 1970s such as the Forest Occupancy Management (1975), the Family Approach to Reforestation (1976), and the Communal Tree Farming (1978). These three forest management programs manifested government's awakening recognition that politically dispensing forest lands to private wealthy individuals through logging rights and concessions can no longer continue, evidenced by the massive forest decline and degradation during this period.

The ISF program granted stewardship agreements to qualified individuals and communities and allowed them to continue to occupy and cultivate upland areas, in return for which they agreed to protect and reforest such land. Security of tenure was for 25 years, renewable for another 25 years through a Certificate of Stewardship (CS) or Certificate of Community Forest Stewardship (CCFS). This first wave of government social forestry projects, however, was primarily rationalized through forest resource protection. People's economic "upliftment" was an unintended by-product, if it did happen. At this point, while government responded to the tenure security concern, social equity issues still remain marginalized.

By the time the Aquino administration took over in 1986, social equity became a core concern in the forestry agenda, accompanied by the new Constitution's provisions on equitable access and benefits of distribution from the country's natural resources. The previous inequities of timber allocations to a favored few came to a close as timber license agreements can no longer be renewed under the new Constitution. The 14-year National Forestation Program (NFP) was launched in 1986, targeting 1.4 million hectares to be reforested by 2000. An ADB/OECF loan of US $240 million in 1988 supported this program further and became the Forestry Sector Project (FSP).

The main objective of reforestation programs before 1987 focused on watershed rehabilitation and protection. The exotic species used in watersheds were the same species used in production forest and private small-scale wood lots. Without data or strategy to ascertain where logs were cut, proclaimed watersheds and protection forests were exposed to illegal cutting while legitimate private production was subjected to costly bureaucratic processes. Local stakeholders including families, communities, corporations, academic institutions, NGOs, and local government were given only short-term contracts with no long-term incentive to sustaining watershed cover.

More recently, the DENR acknowledged the ineffectiveness of such programs. "By and large, past reforestation efforts before 1990 were considered a 'total failure.'"[134] Even after 1990, the quality and effectiveness of reforestation for watershed protection were still in question. The declared function of a watershed often did not match the type of species planted. Reforestation in protected watersheds did not have the adequate social component that responds to thousands of people often seeking livelihood and security within these watersheds.

The problem in protected watersheds has not been adequately addressed but in other sites the Forest Lease Management Agreement (FLMA), a 25-year tenure arrangement, was subsequently introduced. This allowed lease holders who successfully complied with terms of reforestation contracts to harvest, process, sell or otherwise utilize the products grown from the land. In time this became part of a broader forestry program.

A national Community Forestry Program (CFP) was launched in 1989 through Department Administrative Order No. 123 with strong support from Secretary Fulgencio S. Factoran and a diverse group of development agencies and NGOs. From 1990-1995, the program evolved into a national strategy with 12,000 households in fifty-five upland and coastal communities. Based upon a tenure mechanism (the Community Forest Management Agreement or CFMA) providing a renewable twenty-five year lease to qualified community organizations, the CFP program gradually expanded to cover 108 sites and nearly one-quarter of a million hectares. The CFP aimed to provide upland

Plantation species for protection forests did not work. Plantation species were used in enhancing protection forest while production forest has a history in Mindanao of alienating local communities. Traditional reforestation methods were replaced by contract reforestation that provided short-term employment and involved families, communities, corporations, academic institutions, NGOs, and LGU and offered market incentives. MAPALA members planted banana on vacant spaces between acacia mangium.

farmers legal access to forest resources and the financial benefits that can be derived, to create strong incentives for upland residents to implement forest conservation, to establish multi-use forest management by participating communities, and to optimize forest resource utilization and management. According to Marites Danguilan Vitug, who has tracked the Philippine forestry sector for decades, while the CFP program was backed by a broad group of lending agencies, NGOs, academics, foresters, and bureaucrats, and while there was no active opposition to the program, concerns remained regarding the capacity of the Department of Natural Resources (DENR) to implement it. Nonetheless, many perceived CFP to be the best chance to halt further deforestation. During an interview that Vitug held with Secretary Ramos, he noted that: "All the misgivings had apparent solutions. So we viewed the problem as one of fine tuning, rather than the need to re-examine the concept which was accepted virtually unanimously."[135]

Other community forestry efforts mushroomed in the 1990s, such as the Upland Development Project (UDP), the Forest Land Management Program (FLMP), the Low Income Upland Communities Project (LIUCP), the Regional Resources Management Project (RRMP), Integrated Rainforest Management Project (IRMP), the Natural Resource Management Program (NRMP), and the Coastal Environment Program (CEP). The passage of the National Integrated Protected Areas System Act in 1992 emphasized public and community involvement in resource management as the policy encouraged community participation in boundary delineations and protected area management. Protected areas also cover forest lands.

In July 1995, under the leadership of Secretary Victor O. Ramos, the DENR institutionalized the CBFM strategy as the national initiative to attain sustainable forest management and social equity. CBFM integrated all the previous people-oriented forestry programs and new projects were issued a CBFM agreement or CBFMA as the tenure instrument. Project areas covering ancestral domains or lands were issued either a Certificate of Ancestral Domain Claim (CADC) or a Certificate of Ancestral Land Claim (CALC). Similar to previous agreements, CBFMAs and CADCs/CALCs were issued for 25 years, renewable for another 25 years and entitled holders to develop, utilize, and manage specific portions of forest lands according to an approved Community Resource Management Framework (CRMF).

Formal Coverage of CF

By 2004, after seven years of implementation, the Government of the Philippines had "turned-over" 5.9 million hectares to over 4,881 community organizations for management. This includes the CADCs/CALCs issued by DENR covering 2.54 million hectares. This area reflects approximately one-third of the state forest land in the country. What has this entailed? In the Philippines, CBFM refers to a long-term engagement of communities in resource planning, sustainable utilization of forest lands, management plan implementation, and equitable sharing of benefits for the socio-economic improvement of community members. To form a long-term agreement (CBFMA) with the DENR, the community needs to be represented by a registered entity, usually a people's organization (PO) comprised of a group of individuals who can represent the area. The PO members are responsible for site identification, formulation of a 25-year management plan, divided into five-year work plans. Each community is also required to develop and enforce internal resource use rules, and to establish mechanisms and agreements for equitable benefit sharing. The POs are responsible for the protection, rehabilitation and conservation of the natural resources in the CBFM area and to assist government with the protection of adjacent forest lands.

In October 1997, the Indigenous Peoples Rights Act (IPRA) was passed by Congress, laying down the legal framework for recognizing, promoting, and protecting the rights of indigenous peoples (IPs) and indigenous cultural communities. The IPRA provides that development programs, project and activities must be developed along the key elements of the law: recognition and protection of ancestral domain/land rights, self-governance and empowerment, cultural integrity, and social justice and human rights. The IPRA also established the National Commission on Indigenous Peoples (NCIP) to oversee all matters relating to indigenous communities. This meant that the responsibilities for upland areas occupied by the indigenous groups would be transferred from the DENR to the NCIP. Under the new legislation, the NCIP would be responsible for the national program on the delineation of ancestral domain claims through the issuance of Certificates of Ancestral Domain Titles (CADTs) and Certificates of Ancestral Land Titles (CALTs). CADCs previously issued by the DENR under the

CBFM strategy and covering 2.54 million hectares will be converted to CADTs, if the communities so desire.

Clearly, transferring nearly six million hectares of upland area, much of it degraded forests and grasslands (Box 3)[136] to almost five thousand community organizations, is an immense undertaking. Most of this was accomplished in the last five years, and it is not surprising that many of the POs are still in an early stage in developing both their administrative and implementing capacities. Aside from the time required for communities to develop new management skills and institutions, the implementation process was also limited by the ability of the DENR whose efforts were often slow and uncoordinated and fraught with politics. The lack of appropriate training of DENR field personnel on CBFM implementation compounds the limitations of the POs, apathy of some local governments, and inadequate funding. This was compounded by the lack of comprehensive documentation on site location and area that hindered the tracking of systemic problems in an early review of the program and its impact.

Informal Coverage of CF

While CADCs and CFMAs were issued covering millions of hectares in the Philippine uplands, and proscribe national guidelines for complying with the program, some of the area is still managed through indigenous systems of stewardship. Traditional institutional arrangements, tenure instruments, and the customary laws of over one hundred upland cultural systems operate generally at the margins of Philippine society and carry varying degrees of influence over resource management. Where traditional resource management systems remain strong, as they do in the uplands of Northern Luzon, generic CBFM programs may not be supportive of indigenous approaches to resource management.

The Bontoc, for example, have sophisticated traditional systems of land and water management. They are credited for carving the Philippine rice terraces and maintaining these through a *batangan* system that involves protecting selected natural pine and replanting these above the rice terraces, and intensively managing the land for irrigation and food. While the Bontoc are a strong and dynamic cultural community, their way of life is currently under much pressure. Young people seek educational and employment opportunities in the city, where they learn little regarding traditional resource management skills and the ecological values of their elders. Land dispossession is occurring, stemming from age-old tribal boundary conflicts, sometimes in relation to water sources. Government presence and programs in these areas often do not incorporate and draw from the unique cultural and environmental experiences of the Cordillera area. Upland development programs involving reforestation in pine-dominated areas, watershed rehabilitation and protection, sustainable forest management tools and methods, silvicultural practices, agro-forestry and area livelihood options are sometimes found to be inappropriate, manifesting the lack of knowledge and limited appreciation and understanding of the region's culture and environment realities.

Efforts to integrate local knowledge and traditional institutions into national CF programs are under way in some locations in the Philippines. The challenge is to frame the national strategies in such a way that they retain the flexibility to incorporate existing organizations and support existing systems of natural resource stewardship that may have been practiced for generations. Local government officials, DENR staff, NGO facilitators all require special training in order to understand indigenous systems and assess how to relate them to generic CF programs.

While the Philippines has been one of Asia's success stories in rapidly devolving state land management to communities, it faces a number of challenges as it assists communities to manage these resources sustainably. In retrospect, a major weakness in the CBFM policy in the Philippines may be linked to its efforts to include timber production as an important goal of the initial CF strategy. Areas identified for CBFM program implementation emphasized the timber extraction potential of the sites. Less attention was given to the identification of areas for protection, or the conservation responsibilities of the community, even if the area included primary forest. Planners assumed that by setting the cutting rates they could ensure that sustainable use is achieved. In fact, the CBFM strategy was implemented without adequate field capacity to deal with the timber production goals of the program. In many cases, the broader goal of protecting remaining forests was not communicated. CBFM became a result-oriented program with targets allocated for officials to accomplish, which emphasized the signing of a document and not its feasibility nor its sustainability. Further compounding the problem, some CBFM and CADC awards covered exceedingly large

Box 3: Communities and Watershed Governance in Visayas, Philippines

The Carood watershed is a basic example of the ways in which different communities, local government units, and NGOs are attempting to link initiatives to develop integrated resource management systems. Carood Watershed spans portions of six municipalities in eastern Bohol, an island province in the Central Philippines. Covering about 20,500 hectares, approximately 10% percent of which is under forests, some several thousand households live within the watershed. Over the past five years, six Community Forest Management Agreements have been issued covering nearly 2,500 hectares of land, some in the uplands and others for coastal mangrove forests.

The major challenges being faced by villages in the watershed includes unproductive lands (approximately 8,000 ha), wild grass fires in the uplands, soil erosion and downstream flooding in the plains, improper waste disposal, and mangrove cutting, illegal fishing, and coral reef destruction along the coast. Since many of the problems are linked within the watershed, it is critical for local government and communities to collaborate in addressing them. In 1998, as part of the initial implementation of the CBFM program, the Department of Environment and Natural Resources (DENR) established the Ubay-Mabini-Alicia (UMA) Sub-Project. The DENR contracted with a number of local NGOs to implement the project, especially reforestation activities. With assistance from the ESSC, the Carood Watershed Working Group (CWWG) was established to coordinate the efforts of various stakeholders. This reliance on local NGOs for DENR CF project implementation is a common practice in the Philippines, and working groups are frequently used mechanisms to facilitate DENR, NGO, private sector, university, and community interactions.

In the upland communities of Ubay Municipality in 2001, three tree planter's organizations from local barangays were awarded CFMAs and contracts to undertake reforestation. The challenge was to reforest around 500 hectares of cogon (*Imperata cylindrica*) which is highly prone to forest plantation fires. Over a hundred hectares is now planted to forest and agro-forestry species and even fruit trees. The organizations have developed management strategies for fire control including the establishment of fire lines and firefighting training. When the country mapped out its ongoing land use, some of the difficulties became evident. Living along the national highway, the grasslands they replanted are far away to the east and they have to pass through another government program to get there (Map 3).

In the lower watershed, coastal communities have watched the mangrove forests being cleared for fish ponds and their fishery depleted through the use of explosives, which also destroys the fragile coral reefs. To stabilize the environment, local people's organizations have been awarded contracts to restore nearly 800 hectares of mangrove forest. They have also organized a system of protection to guard against illegal fishing practices. The community group is also exploring a range of income-generating activities including mud crab culture, bee-keeping, and eco-tourism.

Grasslands in Upstream Carood. The Department of Agriculture runs the 3,600-hectare Ubay Stock Farm in the headwaters of Carood. Residents from Barangay Bongbong were paid to plant a portion of this area with gmelina under a DENR watershed reforestation project. They had to walk two kilometers uphill and cross two of the Stock Farm's fences to get to the project's designated reforestation area. This later became a CBFM site, but so far tenure has not yet been issued to Bongbong residents.

While the CFMA program in the Carood Watershed has established agreements with six community groups, through local people's organizations, funded their reforestation and management activities, and provided some technical assistance and training programs, the program also faces a number of ongoing challenges. These issues range from political conflicts between rebel groups and the military in the uplands, to illegal economic activities such as dynamite fishing along the coast. Yet, the most critical problem is jurisdictional coor-

MAP 3. Community Resource Map, Bongbong Village, Bohol, Philippines

dination between government agencies, as well as between local government units. Such coordination is critically important and strategic in integrating NRM project activities within a single watershed. Even in a small watershed like Carood, linking different administrative institutions and development agencies is a difficult task. Community forest management activities are also usually dependent on external project funding, and once such support ends, sustaining local initiatives can be problematic if communities are not already empowered and there are no actual benefits. Efforts by ESSC to facilitate the development of an integrated watershed management plan for the Carood through the Carood Working Group have been received well by a number of the municipal governments and people's organizations, however, the integration of community-level management systems at a watershed level have been slower to develop.

As a response to these and the concerns also raised by local government representatives, the Working Group decided to create a formal body that could effectively respond and plan proactively for the management of the watershed. Initial discussions led to the formation of the Carood Watershed Management Council created through a verbal commitment among the members and later through an Executive Order of the provincial governor. The members signed a Memorandum of Understanding to formalize their agreement and facilitate the accessing of resources for the sustainability of the management body's activities. The Council has acknowledged the need to harmonize LGUs' development plans and communities' resource management plans, and to integrate their implementation with the assistance and cooperation of all stakeholders to ensure the sustainable management of the watershed.

A recent review of CFM in Bohol,[132] identified both strengths and weakness of CFM that are shared in many other provinces around the country. It was noted that CFM works better in areas that have comprehensive programs for poverty reduction. The convergence of non-government agencies, local government units and non-government organizations (NGA-LGU-NGO) support programs in CFM areas enhances resource management, lessens dependence on forest resources for a living, and overall, improves the quality of life of the community. CBRMP helps promotes agroforestry by distributing fruit tree seedlings and livestock.

Mangrove Resources in Lower Carood. In contrast to its resource-poor uplands with 68 hectares of secondary growth forest, downstream Carood has 291 hectares of mangrove forest. With over 30 species, the mangroves in Carood is considered a healthy ecosystem able to support local livelihood.

CFM works well in areas with harmonious clan relations, where the traditional Boholano culture of voluntary cooperation still exists. CFM also works when the barangay LGU plays a frontline role. The tasks of forest protection, firefighting, and building of barangay infrastructure, is anchored on community mobilization – a voluntary collective action weaved by traditional bonds and driven by common interests. This pattern is true to many well-managed CFM areas in Bohol and reveals that the socio-cultural aspect is an important consideration in planning and implementation of community forest management. Ecotourism is an emerging strategy for resource conservation and livelihood development in Bohol's CFM areas. Many of the assessed CFM sites have potential for ecotourism and a few have operational tourism projects. Community-managed ecotourism includes recreational activities (nature trek, river cruise, bird watching and cave exploration), selling of goods and services (rowing, tour guiding, massage, etc.). Tourism boosts livelihood activities, and generates employment and income for the community. Tourism centers can be complemented with satellite ecotourism sites in CFM areas. In this way, livelihood opportunities are broadened and conservation of natural resources is reinforced.

On the other hand, it has been found that the mere processing and installation of the CBFM Agreement alone is inadequate. In most cases, the CBFMA did not undergo the process of orientation and negotiation with stakeholders. Thus the community cannot develop self-determination and self-reliance – both basic principles of CFM. The Community Resource Management Framework (CRMF) does not comply with the minimum requirements of DENR guidelines. Documents are usually prepared by DENR with insignificant participation of the community. Moreover, CRMFs are often not used or referred to by the CBFMA holder. The ill effects are now taking its toll on CFM. POs are unable to cope with day-to-day decisions needed for CFM implementation without DENR intervention. They remain dependent on external institutions such as government agencies and NGOs for leadership and management of the CBFM.

Perhaps one of the biggest challenges facing community forestry in the Philippines in coming decades is to successfully enhance livelihoods through increased forest-based income flows. After over a decade of field projects, resource management in many CFM areas has not produced the "promised" economic benefits and communities remain poor. Those contracted for reforestation projects, have long been waiting for the issuance of the Resource Use Permit which is necessary so that the community may utilize the resources. Once granted, however, the community must contend with identifying and linking up with markets. ■

areas, beyond the capacities of communities to effectively manage. Even where communities were becoming productive, the clearance procedures and access to markets were poorly developed but for selected sites.

Problems emerging from an emphasis on timber production in early community forest management programs are not uncommon in other countries. Typically, communities have managed forests in the past for multiple purposes including subsistence fuel and timber, NTFPs, and a wide range of environmental services. Most Asian communities have little or no experience with commercial timber harvesting and marketing. Programs that emphasize timber production frequently do so due to the needs of funding organizations to financially legitimize their loans by demonstrating income flows. At the same time, the local private sector and some government actors may be interested in continuing logging activities after other forms of commercial felling have expired or have been halted due to past abuse or policy changes. The commercial potentials developed with migrant communities in these areas were in many cases too closely aligned with large-scale production.

In the Philippines, the community-based timber production orientation of CBFM and CADC, were de-emphasized during the term of Elisea G. Gozun, DENR Secretary from 2003-2004. She noted that the primary aim of CBFM was to sustain the remaining forests and not sustain wood production. Forests were de-emphasized as commodities and a major element in the national economy, and given greater importance to their role in providing social stability and a basis for poverty alleviation. DENR policy makers acknowledged that the CBFM program needed to ensure a level of harvesting in some areas to sustain community's livelihood and small enterprise requirements. The Philippines went through an initial review of the CBFM program goals and CF area allocations in order to: (1) cancel corrupt or misallocated resource use agreements, (2) acknowledge where and why CADC/CBFM agreements are not working and, (3) support a greater level of community self-monitoring. This required most management plans to be rewritten.

In 2005 under the new term of President Arroyo, the DENR Secretary changed again. Under the new administration of Michael Defensor, the CBFM assessment has been compounded by renewed questioning without the DENR first acting upon existing recommendations and attaining critical finances for implementation. Communities as well as local line agency officials are confused as to what to do. On top of this, people perceive new incentives for pulp plantation investment as threats to the secondary forests in areas being allocated to corporations. Finally, the series of policy actions due to the floods in December 2004 have sent mixed signals to local governments, NGOs, and communities as to the forest management agenda of DENR. A log ban was issued one month after the floods only to be lifted after a few months, with even an attempt to reinstate commercial logging in areas formerly out of bounds. People are waiting for government to clarify its unwavering commitment to CBFM. Transparency and accountability in DENR activities are again dissipating, yet the trend towards greater devolution of forest management rights will likely be sustained.

The problem is no longer a question of achieving CF coverage for a target area, but rather responding to needs in site-specific contexts and ensuring effective forest management in each locale. A major challenge to the achievement of this goal is the diversity present in the Philippines, not just in terms of ecological and cultural conditions, but in terms the types of pressures on the resource, and varying levels of capacity to respond to them. Land insecurity stems not only from inadequate resource tenure, but also from a lack of basic services and appropriate skills, limited trust in local government, social conflicts among stakeholders, and poor utilization of resources and markets.

Community forest management was adopted as a leading, long-term national strategy in the Philippines. While its implementation has accelerated or decelerated under the past four or five administrations, it was never questioned as a key element in natural resource management. In terms of social justice and environmental action, community forest management has become the only politically acceptable strategy for most public lands management. The extent to which CBFM will be successful in restoring environmental health will not be determined for some years, and possibly decades, to come. Natural forest regeneration and reforestation efforts will likely be slow due to the extent of ecological degradation in much of the uplands, and the gradual development of community capacities to implement management activities. Yet, for the most part, lands transferred to POs are increasingly being protected, and much of the territory is showing signs of ecological recovery, even if such resources are yet to provide significant livelihood.

CHAPTER 10

Thailand

Like many nations in Southeast Asia, Thailand has a diversity of indigenous and spontaneous forms of forest management. Despite rapid social change and modernization, community forest management systems continue to operate around the country, with some based on more traditional modes of stewardship while others are rapidly adapting to respond to the dynamic economic environment. Thailand was one of the first nations in the region to experiment with participatory forest management schemes in the early 1970s, however, progress in developing a legal and policy framework to strengthen community forest tenure rights has been slow. As a consequence, a nationally ratified and organized community forest management movement has yet to take shape on the ground.

To understand why the impact of formal CF programs has been slow to develop in Thailand it is useful to review the recent history of the sector. Thomas and Pragtong view this development in four major phases.[137] During the first phase, the Royal Forest Department was created in 1896 in an effort to control the growing trade in timber, especially teak, and to capture greater revenue for the emerging nation state. In 1899, forest ownership was formally claimed by the central government. Until the mid-1950s Thailand continued to focus on revenue generation from forest timber. The RFD viewed communities primarily as a source of labor and populations to be regulated, though some accommodations were made for subsistence use. From 1954 through 1967, the policy focus shifted from timber production to forest land allocation for economic development. In the 1980s and 1990s, when large tracts of forest lands were allocated for national parks, wildlife sanctuaries, and watershed conservation areas, land conflicts with communities became more common. According to one report:

> *RFD officials realized that reserved areas included some existing villages, but they assumed procedures would be established for processing claims to previously held lands. In mountainous areas, however, many ethnic minorities had no legal status in Thai society, and claims to lands under shifting cultivation and forest fallow were difficult to fit under lowland Thai Land legislation.*[138]

Between 1954 and 1967, forest cover in Thailand declined from 60 per cent to 48 percent. In 1968, the national government's push for economic development resulted in more than 500 new timber concession areas declared and by 1980 forest cover had dropped to 36 per cent. While industrial timber production was an emphasis during the 3rd Phase of RFD's development, the agency also began a Forest Village Program with the Watershed Management Division and thereby began gaining experience with community planning and rural development projects. During the 4th Phase, the Royal Forest Department's policies became increasingly shaped by the national social and environmental debate. In-migration into the northeast by landless people from the central regions resulted in a rapid loss of forest cover. In 1988, however, the public dramatically rejected industrial logging policies, after a wave of floods and landslides occurred in November with 200 dead. Public outrage triggered by media coverage and publicity generated by environmental groups resulted in a logging ban in reserved forests and the revoking of all timber concessions in 1989. These events, for the most part, marked the end of industrial logging in Thailand.

In recent years, community forest policy has been hotly debated at the highest level of Thai government. Both the lower and upper houses voted on a CFM bill that has evolved over the past decade through a broad-based civil society dialogue involving political leadership, foresters, academics, environmental NGOs, and NGOs working on social development issues. The policy under review is important, as it has the potential to impact the lives of millions of rural people and change management authority over a significant part of Thailand's total land area. Passing through the Lower House unchanged, the bill met serious resistance in the Thai Senate, where key clauses were gutted to the great disappointment of the bill's advocates.

This story begins after World War II, when Thailand was largely a rural, agrarian nation, with nearly two-thirds of its land covered by evergreen and dry deciduous forests. Thailand has changed dramatically since that time. Its population has increased from 23 million in 1961, to 62 million, just forty years later. While Thailand has experienced rapid economic growth and urbanization over the period, over three-quarters of its population still resides in rural areas and many retain some subsistence dependence on forest lands.

National forest cover in Thailand has fallen at almost the same rate that the population has risen. In 1953, 60 percent of the country was estimated to be covered by forest, while by 2001, that figure had fallen to 29 percent according to FAO statistics, though many experts consider this to be high and actual forest cover to be around 18 percent if plantations are excluded.[139] Forest clearing in Thailand has been driven by two related and often sequential factors: commercial logging, and clearing for agriculture. The logging and clearing of forest land for agriculture has been most dramatic in the northeastern part of the country, where approximately 50 percent of all forests were cleared between 1973 and 1982.[140]

Throughout the 1970s, fears of communist insurgency in the northeast and desire to expand cassava exports resulted in a lax attitude towards forest encroachment with much of the northeastern forests cleared for agriculture. This was reflected in the forest land allocation policy that provided land tenure certificates to households occupying degraded reserved forest areas before 1982.[141] Thai timber companies, with strong ties to high-level military and political figures, rapidly depleted public forests with commercially valuable hardwoods. The dramatic decline in forests made Bangkok headlines in 1989, when torrential floods inundated the city. Public outrage from forest mismanagement and corruption led to a commercial logging ban later that year. The end of the industrial logging era in Thailand created an expectation among some NGOs that new opportunities for community-based management[142] might be possible. The logging ban also shifted logging pressures to neighboring Burma, Cambodia, and Laos.

Decentralization and Constitutional Reform

Along with other nations in Southeast Asia, Thailand embarked on a decentralization program in the early 1990s during a period of rapid economic growth and liberalization.[143] The *Tambon* Administration Act of 1994 established Tambon Administrative Organizations (TAOs) in 7400 sub-districts nationwide, delegating responsibility for a range of development planning, implementation, and fiscal management duties to local government officials. Aside from education, infrastructure, health and welfare, natural resource management responsibilities and rights were also included. While the TAO Act strengthened the role of local government in resource planning and use, it also conflicted with emerging policies governing conservation areas that were managed through centralized line agencies.

The Tambon Administrative Act was further empowered by the 1997 Constitution, which also mandates that sub-district (*tambon*) councils of the village representative (*obator*) would have a greater allocation of local government budgets for resource management and community development. This shift towards devolving resource management authority to local government was further extended in 2002, when the Watershed Management Unit of the Royal Forest Department (RFD), responsible for the vast majority of northern Thai forests, was placed under the Ministry of Interior. While the growing authority of the Ministry of Interior over forests might be seen as signaling greater decentralization of authority, strong environmental policies in Thailand have required the Ministry of Interior to exert strong central control over many forest areas and watersheds, especially in the north, and in some cases this has diminished community forest access rights.

While the laws and policies dealing with governance have been important in redefining the rights and responsibilities of communities for local forests, the Royal Forest Department (RFD) has taken a mixed position regarding formalizing and supporting community involvement. While a conservative element within the RFD has consistently opposed such initiatives, a growing number of younger, progressive foresters have moved to design and implement collaborative management systems throughout the community. A community forestry unit was established in the early 1990s, that later grew into a division with several hundred staff. In 1992, this unit was responsible for a national inventory that documented over 12,000 rural groups protecting forests ranging in size from 1 to 4,000 hectares.[144] They have also been responsible for organizing numerous community forestry training programs for RFD staff in an effort to reorient the national forestry agency. Yet, despite a decade of struggle, dialogue with NGOs, academics, and political leadership, they have not had the political influence to drive the passing of national CF programs or policies.

While the 1997 Constitution enshrines the rights of communities to local resources, it requires the implementing guidelines that were anticipated to be forthcoming through the Community Forestry Bill.

Community Forestry Bill

The community forestry bill is the most recent outcome of a long debate over the role of communities in managing public forest lands in Thailand. The first draft of CF legislation was done by the RFD in 1992, but was limited to addressing the role of communities with fast-growing tree plantations. In 1993, an NGO group (NGO-COD), comprised of people's organizations and academics, began lobbying for greater recognition of community rights to manage local forests. In 1995, Prime Minister Bunharn's cabinet endorsed a policy to allow communities to reside in conservation areas, but this initiative was vigorously opposed by a group of environmental NGOs. During the administration of Prime Minister Yongjaiyuth, a further legislative initiative was launched to create community forests managed for commercial purposes, and while the draft bill was endorsed by the cabinet, it was rejected by people's organizations and social NGOs. The Prime Minister tried again in 1999 with a revised draft bill that recognized community forests in production areas, but excluded them from any management role in conservation forests. This bill stalled while under review by the RFD and the Ministry of Agriculture and Cooperative.

In 2000, a people's coalition with nearly 53,000 signatories submitted a new CF bill to the Thai Congress emphasizing the rights of local communities to manage all natural resources through a multi-sector committee, based on a five-year demonstration of competence. The CF bill was accepted by a parliamentary commission, along with draft CF bills from four other parties and one from the government. The Parliament was dissolved before it had a chance to conclude its hearings concerning the draft bills. The government's draft CF bill, however, was taken up again by the Rak Thai Government in late 2000 and endorsed by the Lower House in November 2001 by an unanimous vote.[145] The Lower House forwarded the Community Forestry Bill to the Senate for final approval. The bill ran into stiff opposition, however, as the amendments imposed by the Senate stripped the legislation of a number of important rights and privileges that it would have conveyed to forest-dependent communities. Some of the key modifications are described in Table 9.

The Senate amendments to Article 18 caused serious concerns for CF bill supporters, as they excluded smaller hamlets from recognition, as well as communities living within protected areas and recently established CF groups. Changes to Article 29 also undermined the intent of the original CF bill by reducing flexibility in the management transition process and the sharing of authority between the Royal Forest Department and the community. Past experience indicated that communities and forest field staff may extend the CF territory as the community develops management systems and acquires the capital and technical resources to implement their plans. This allows for the gradual expansion of CF areas until all relevant local forests come under community stewardship. It also permits

TABLE 9: Key Elements in Thailand's Community Forestry Bill and Proposed Changes

Article No.	Original CFM Bill	Senate Amendment
18	CFM tenure for at least 50 persons above 18 years of age from traditional groups native to the area and actively engaged in forest preservation for at least the previous five years.	– Raises number of proponents to 100 persons above 18 years of age, – Excludes protected forests including watershed and wildlife sanctuaries, – Limits timeframe to five years "before the bill takes effect."
29	Allows community forest groups to request changes in the boundaries of community forest area in order to improve the management plan.	– Prohibits the expansion of a designated community forest area.
31	Trees may be cut only for subsistence and public utility, which should follow guidelines set by a relevant policy committee.	– Prohibits collection of any forest products in the community forest, unless prior permission is secured from the Forestry Department.

community groups that find they are unable to manage the lands, to turn over those responsibilities to neighboring groups, or to redefine boundaries in ways that allow for a more efficient division of territorial responsibilities. The amendment effectively eliminated flexibility to make adjustments in forest areas assigned.

A fundamental goal of most CF policies is to secure community rights to forest products. The original CF bill had already restricted utilization to subsistence requirements. The amendment required communities to seek permission from the RFD for extraction, a stipulation that in other countries has resulted in "illicit use" and corruption and constrained the ability of communities to actively manage forests. As one villager noted, "Do we have to ask a forestry official every time to pick a mushroom from the village floor? This goes against our way of life."[146]

It is helpful to examine reasons why 157 of the Thai Senators resisted approving the original Community Forestry Bill. The reasons differ widely. While a significant amount of resistance may have come from individuals representing more conventional views of the forest department and private sector, it is apparent that many liberals and environmental groups were also in opposition. According to news reports appearing in March 2002, "Some said they were afraid that if the villagers received rights to manage the forest, they would convert the fertile forests to cabbage and other cash crops. Some were afraid 'outsiders' might abuse the bill by encroaching on protected forest and claim rights to manage it as community forest."[147]

Some Senators wanted the bill to be tested on encroached forests, others wanted it restricted to degraded forest land, while others rejected it outright as a threat to protected forest areas. As Senator Thongbai, a renowned human-rights advocate noted, "For the present, they (forest dwelling communities) behave well in managing the forest, but in the next 10 years when their community grows, how can they survive if they don't encroach on more forest areas?"[148]

Encroachment is certainly a major concern raised by those against the bill. One opposing Senator claimed that article 18 was a ploy to give villagers special privileges to stay and make a living in protected forests which were for future generations. "They have run their businesses in protected forest – how could they ask the right to manage the plots? Articles 18 would give them the legal right to claim their cabbage plantations as community forest."[149] Yet, the supporters of the bill claimed that many of the senators in opposition, whether liberals or conservatives, environmentalists or human rights advocates, were largely upper class urbanites who had little understanding regarding the realities of the nation's forest dwelling peoples.

One senate representative from Ubon Ratchatani, Dr. Niran Pittakwatchara, is a supporter of the bill and sees it as a logical approach to sustaining forest resources. He observes, "We should admit that the Forestry Department doesn't have enough resources to protect all of our forests...The problem is that many areas where locals have already been managing the forests have been declared national parks and other forestry classifications." He acknowledges that Senate rejection of the bill reflects a lack of information and attitudes. "If people know, they will see how the draft bill is crucial for forest management in Thailand."[150] Another supporting Senator, Chirmsak Pinthong, remarked, "If I had never studied forests and communities, I might have voted against Article 18 too." Chirmsak organized field trips to allow senators to see how effectively some communities were already managing protected forest lands, but found only those who already supported the bill joined the trip. He insists that the bill has ample protection to ensure that new encroachments would be stopped.

Pearmsak Makarabhirom, formerly of the Regional Community Forestry Training Centre (RECOFTC) noted, "For villagers, the most important things that will save the forest are their culture, traditions, and beliefs. Their culture respects the forest, but is never reflected in any forest protection laws drawn up by people whose lives are in urban areas."[151] He noted he was not surprised that legal experts hold different views and the law never accepts culture as a social principle, rather that only written regulations under legislative power are recognized as law. Nonetheless, he feels cultural rules are important tools for society to live in harmony with nature.

The new Senate version of the CF bill returned to the Lower House, which may either support the new version and bring it into law, or reject it. If the Lower House insists on backing the original version, a joint Senate-House panel will be established to discuss the bill further. Political resistance to reforming the forestry sector and recognizing community rights and responsibilities to manage local resources through CF legislation has frustrated those who advocate change. The strategy of basing a CF bill on constitutional articles ran into opposition from too many different interest groups. Conflicts within the Thai NGO community, particularly between the "dark green" environmental NGOs, and the "light green" social NGOs have created a polarization of positions on the CF bill that have created serious impediments to compromise.

CF advocates have recently adopted another approach to formalizing community forest management through bureaucratic reform and stronger linkages to decentralization and devolution policies. In October 2002, the Watershed Division of the Royal Forest Department, that has sponsored numerous CF pilot projects and is responsible for the field level management of public forest lands in much of northern Thailand, was transferred to the Ministry of Interior. This change means that sub-watershed unit chiefs in the field will no longer be under the direct supervision of the RFD in Bangkok, but will be taking instructions from local government.

Thailand continues to suffer from an absence of comprehensive legislation dealing with the resource rights and responsibilities of forest-dependent and resident populations, many of whom are ethnic minorities. As recently as February 2005, the Permanent Secretary of the Ministry of Natural Resources and the Environment, Mr. Petitpong Pungbun Na Ayudhya, noted that villagers should be encouraged to play a larger role in forest protection and conservation.[152] Yet, while many senior government planners endorse the CF Bill, the process of CF law formulation has been stalled for over a decade by inherent disagreements among key stakeholders including environmental lobbyists who fear that empowering community forest management would lead to further forest clearing. This stalemate has created a legal and policy environment where there are conflicting programs and agendas fostering confusion and lack of action on the ground. Forest loss in Thailand is estimated between 60,000 to 160,000 hectares annually.

National Program Implementation

In 1986, a community forestry section was created within the RFD to spearhead the development of new participatory programs. A growing number of Thai NGOs and academics were also developing expertise in community forestry programs and strategies at the same time, some of whom had begun to work closely with the CF Office of the RFD. Unfortunately, many RFD staff remained skeptical regarding a larger role for community resource management, especially in the northern uplands. One RFD official estimates that not more than 10 percent of officers were knowledgeable or positive concerning devolving greater management rights to communities in the late 1980s.[153] The Community Forestry Office also lacked the financing and staff to launch a major national program. In fact, the entire full-time staff of the RFD did not exceed 12,000 persons during this period, many of them based in Bangkok, provincial, and district centers. Equally constraining, a national CF policy did not exist to guide a major initiative, and consequently most efforts existed as scattered pilot projects funded by different development agencies. As discussed in Part III, however, a long-term effort to form a national CF policy was already under way, and continues to the present time.

Official Coverage of CF

In Thailand, a national inventory conducted by the Royal Forest Department in 1992 documented over 12,000 rural groups protecting forest patches ranging in size from as few as one hectare as many as 4,000 hectares.[154] Some of this territory is privately held, while other community forestry areas operate informally on lands that are legally part of state domain. It is difficult to estimate with any certainty what area could be considered legal or official community forestry areas. The problem stems from a lack of clear definition of where community forest management can be formally approved and registered by the government, and what forms it can take. The absence of a CF policy is not for lack of thoughtful debate. The national CF bill has been under development for over a decade, with much input from a wide range of stakeholders (Part III). The absence of the CF bill has imposed major barriers on NGOs and Royal Forest Department staff who have been attempting to formally engaged communities in forest protection and management for over twenty years.

Informal Coverage of CF

Community forestry takes many forms and serves different functions in Thailand. In the South, forest lands are frequently under the ownership of private individuals, though community management systems are expanding in some mangrove forests along the coast.

While the RFD had leased these areas for fuelwood harvesting to communities in past decades, a growing number of villages are forming management committees to better protect and regenerate them, and prevent them from being converted into aquaculture and other commercial uses. In the northeast, much of the forests was lost to agricultural expansion in the 1970s and 1980s. Due to the massive in-migration occurring during that period, many northeastern communities are comprised of recent arrivals from different parts of the country, and as a consequence there may be no local traditions of community forest management.

Nationwide, at least four major types of community forestry can be identified. These include 1) newly organized community protected forests that have emerged as a response to illegal logging; 2) monastery (*wat*) forests that are restricted areas (plants and animals are protected), 3) wetland forests that communities protect to ensure there is a breeding ground for fish, frogs, and crabs, as well as a source of bamboo, timber, and fuelwood, and 4) cultural forests that have economic, historical or religious significance. In some northeastern cases, CF systems are networked such as the 4,000-hectare dry *Dipterocarpus* forest of Dong Yai that is protected by 12 villages.[155] In the case of Dong Yai, the community began encouraging forest regeneration after the commercial *kenaf* market crashed and newly cleared forest land began to send up shoots. The RFD cancelled logging concessions in 1989, and since that time the forest has recovered rapidly under community protection. It may be that much of the extensive area of forest cleared for commercial cassava cultivation thirty years ago will eventually be reforested, following the experience of Dong Yai.

As in most countries in Asia, it is difficult to calculate the area informally under community management in Thailand. One Thai NGO estimates that in northern Thailand, traditional systems may cover 328,000 hectares or more.[156] There may be another 1,000 community forests in the northeastern part of the country. Some of the forests that have been managed by indigenous communities, or taken over by migrants more recently, have been formally recognized under the STK Program which issues land usufruct certificates to households. Since 1982, the RFD has surveyed most of the nation's forests and has issued over 800,000 land use certificates to over 700,000 households covering 1,150,000 hectares or approximately two percent of Thailand's land area, or four percent of the official forest area.[157] This certification process, however, focuses on forest land that has been permanently converted to agriculture and under household management, rather than forest areas under communal stewardship.

Thailand has taken a unique position among its Southeast Asian neighbors in banning commercial timber concessions 15 years ago and generally halting logging in natural forests. Levels of illegal logging are, quite probably, lower than any other country in the region. One reason for the success of this moratorium on industrial timber felling on state forest lands relates to the strong commitment of the middle-class to protect the natural environment. It is also due to the ability of Thai industry to draw timber supplies from Burma, Cambodia and Laos. Unlike the Philippines, Indonesia, and Laos, where community forestry often emphasized commercial logging, this has not been the case in Thailand, at least not since the early 1990s. Instead, CF type programs have focused on dealing with land tenure issues, NTFPs, plantations, and agroforestry.

A reflection of the trend towards natural resource conservation in Thai society was the transfer of the former Watershed Management Division from the Royal Forest Department to the National Parks, Wildlife, and Plant Conservation under the newly created Ministry of Natural Resources and Environment in 2003. The renamed Watershed Conservation and Management (WCMO), has not changed its mission, however, and has implemented community-based approaches to forest management in over 600 villages in northern Thailand. WCMO has formed a working group to improve coordination with the Department of Local Administration and the Worldwide Fund for Nature to pursue community-local government dialogues.

It is also important to note that community resource managers are developing regional networks to share learning and better represent their views to national government policy makers. In the late 1980s, in response to the growing number of communities under threat of resettlement, a number of northern Thai NGOs formed the Northern Development Foundation (NDF) that has provided input into the development of the community forestry bill and has supported the formation of the Northern Farmer's Network (NFN), an association that links 107 villages located in 14 sub-watersheds. Many fall within the boundaries of 19

Box 4: Communities for Watershed Protection in Mae Khan, Thailand

The Mae Khan Sub-watershed is an informative example of the ways in which the Watershed Conservation and Management Office (WCMO) has attempted to enable and formalize community forest management while awaiting appropriate national policies. In the Mae Khan case, the strategy has involved working with decentralization policies and programs to link upland and highland communities through local government institutions.

The Tambon Administrative Organization (TAO) is a sub-district level body of elected representatives that are charged under existing laws to formulate development and natural resource management plans. Since the late 1990s, the TAOs have been allocated funds for such forest management tasks as fire control, reforestation, and road maintenance. The goal of the Mae Khan project was to build the capacity of village groups, drawing on indigenous leaders and institutions, to develop micro-watershed plans and to integrate them and approve them at the Tambon level through a series of meetings, workshops, cross visits, and planning activities.

The RFD has been pursuing community-oriented watershed management in the north through a variety of projects since 1981. The Sam Mun Highland Development Project (SM-HDP) funded by UNDCP was particularly important in developing collaborative resource planning process and pioneered the use of three-dimensional models for micro and sub-watersheds. Constructed by community members with support from NGOs and RFD staff, these tools provided a framework for discussing forest and land management issues and conflicts. Since that time, participatory land use-planning (PLUP) has become widely used by the WCMO.

The Mae Khan Sub-Watershed is part of the Ping basin, one of six major river systems that flow through northern Thailand. The Mae Khan is situated between 270 meters at its outlet to over 2,000 meters along the ridge tops. Four major ethnic groups inhabit the area with the Hmong and Lisu in highland settlements, the Karen in the midlands, and the Thai in the lower watershed. The Karen and Hmong populations have inhabited the area for around one hundred years, but since the region was declared a protected watershed forest, their land claims have been thrown into question. National conservation-oriented policies have strong national support, especially among the urban middle-class, who represent a large and influential component of Thailand's electorate. Sentiment against forest resident communities is also based on ethnic perceptions of Thais and non-Thais many of whom are of a different kingdom of old or are migrants over recent centuries from southern China and present neighboring countries. In the sub-watershed of the Mae Khan, two Karen villages have an intensive rice-based livelihood while utilizing the neighboring forest. RFD has mapped with them their resource use and maintains a GIS database to monitor their land use in keeping with the forestry policies (Map 4).

Hmong Vegetable Plots. The Hmong in Mae Khan practice shifting cultivation growing upland rice, vegetables, corn on cleared slopes in the upper watershed.

Hill tribes are sometimes perceived as "forest eaters," and some Thai environmental organizations have strongly lobbied for their removal from the forests in upland watersheds in the north.[155] The presence of the highly valued Doi Inthanon National Park on the southern edge of the Mae Khan Watershed further reinforces national interest in conservation in the area. Other organizations, however, have worked in the Mae Khan for over a decade to find ways to integrate local communities into formal watershed management systems. These include both local NGOs operating out of Chiang Mai who support the community development and a networking of hill tribes, Royal Projects, as well as the Watershed Management Unit (WMU) of the Royal Forest Department.

MAP 4. Resource Uses, Ban Pa Gluay—Pong Lomraeng, Mae Khan Thailand

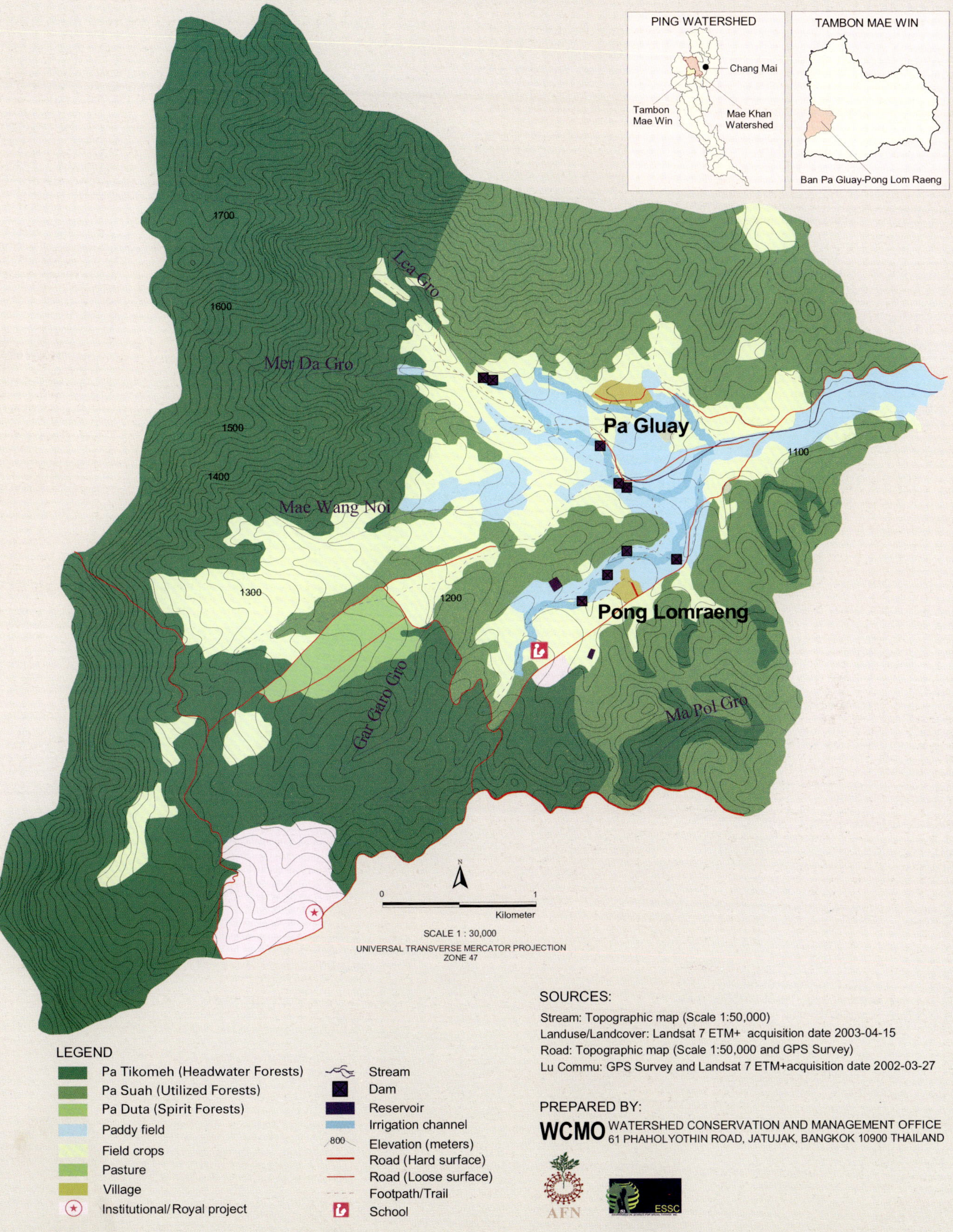

In the Mae Khan each ethnic group has well-established traditions of forest protection and management. Each cultural community has different names, use practices, and regulations for their protection and utilization forests. Despite the accusations of some organizations that upland and highland communities are the destroyers of northern watersheds, much of the damage to forests of the Mae Khan occurred in the 1970s after an access road was constructed and logging concessions began operating in the area. The environmental impact of logging operations and forest clearing resulted in flooding and landslides and spurred local communities to strengthen resource use controls and regulations. Over the past ten years, many of the 21 hamlets in the Mae Khan Sub-Watershed, with support of the WMU and NGOs, have elaborated and codified their forest management rules, mapping and zoned their areas, and formalized their resource use committees.

The integration of three different cultural strategies of resource stewardship has been one of the biggest challenges for the pilot project. A major goal of the project was to formally integrate indigenous forest management systems into the Tambon natural resource management plan, while allowing them to maintain their individual ethnic characteristics, including leadership patterns, values and belief structures, and local terminologies. According to the TAO Act (an important component of Thailand's decentralization laws), is that two members from each village are elected to hold office and serve on the Tambon council (*obator*). Three regular meetings are held each year to discuss development and natural resource management activities. A major topic is fire control and includes the planning and implementation of fire breaks throughout the watershed. Each village is responsible for constructing a 10 to 20 kilometer long fire break drawing on community labor, with modest financial support ($70) from the Tambon. Forest patrolling and tree planting are also activities that are planned through the Tambon council and implemented by each community.

Through the Tambon council, and with the facilitation of the WMU project, a number of NRM issues are gradually being addressed. Conflict between the Hmong and the Karen over land rights and upstream pesticide contamination are gradually being resolved, while boundary disputes with Doi Inthanon National Park and the Hmong community are under discussion. Tension between the Karen and the Hmong and Lisu, who migrated into the area more recently, is often based on the Karen's complaint that the newer inhabitants do not respect customary land tenure.[156] Participatory land use planning processes and coordination at the Tambon level appears to be an effective way to address such problems. The longer-term question of the tenure status of upland and highland communities continues to await national legislative decisions; however, the commitment of local government and technical agencies to support participatory resource planning with the hill tribe communities has lessened anxiety and provided a basis for collaboration between Hmong, Karen, and Thai villages. ■

Karen Rice Paddies. The Karen traditionally cultivates rice on terraces along narrow valleys close to settlement and streams and uses indigenous irrigation systems. The Karen also have swidden fields located further from settlements, with fallow periods sufficiently long enough to allow natural regeneration.

established or proposed protect areas. All NFN member communities possess forest management committees and promulgate forest use regulations in addition to managing community forest funds through annual household fees. Representatives from watershed networks participate in NFN regional monthly meetings. According to the NFN, member communities play an important role in managing the hydrological systems that benefit both upland and lowland communities.

> *It is important that people living in each area have their own watershed forest. Forests in highland watersheds that hill people have taken care of might be able to supply water needed now in both the agricultural areas of the highland and lowlands. But it certainly cannot keep pace with the ever-increasing demands for water and forest made by the lowlanders. What we need is cooperation between people to protect the forests so there is a mutual benefit in the recovery of forest in both the lowland and highlands that helps keep all watersheds in balance.*[158]

With 55 percent of Thailand's forests, much of it with good forest cover, the north will continue to be a primary application area for CF in the future. The fact that forest resident communities are often hill tribes with communal tenure systems and traditions of forest stewardship, make community forestry an appropriate approach for this region. While conflicts over water and political pressure to resettle forest residents will continue to create anxieties among villagers in the northern Class 1A watershed. Watershed Class 1A includes headwaters of rivers, usually located at higher elevation and steep slopes, and should remain as permanent forest cover where no human disturbance is allowed. CF has strong support among the communities themselves, from most academics, and a majority of development oriented NGOs. Decentralization policies and participatory planning exercises supported by government agencies are normalizing community management systems throughout the north, enhancing forest protection, stabilizing forest use, and helping ease resource conflicts. The case of the Mae Khan Sub-watershed reveals how such policies are shaping approaches to forest management in ways that brings local government and hamlet representatives together in a common effort to resolve conflicts, address external pressures, and enhance their resource rights and tenure security (Box 4[159]) [160]. While decentralization policies alone will not replace clear and enabling community forestry legislation, it provides a platform for forestry agencies and local government to work with upland communities.

Due to Thailand's diversity, community forestry programs are taking different forms to effectively respond to a variety of socio-cultural, economic, political and environmental contexts. The ethnic hill tribes that reside in the northern uplands and highlands continue to be dependent on forest resources, yet they live in areas that have been set aside for strict conservation by the state. While the ethnic minorities, as well as the Thai communities residing in the watersheds, all have ongoing traditions of forest protection, some elements of the government and Thai civil society would prefer to have removed resident populations from much of the northern watersheds. The community forestry bill, by default, has become a strategic piece of legislation that may determine their tenurial fate. Its acceptance or rejection will help determine what role resident people will be allowed to play in public forest lands management.

Throughout Thailand, there are an estimated seven million hectares of degraded state forest land, much of it inhabited. If a truly enabling community forestry bill can be approved by the Thai legislature, it is likely CF will rapidly be integrated as a major component in the forestry sector. However, until then it will continue to be implemented through decentralization programs and related community development initiatives.

CHAPTER II

Vietnam

Vietnam nationalized much of its midland and upland forest in the northern part of the country in the late 1950s and early '60s, extending state management to the southern forests after reunification in the mid-1970s. Two major policies that dominated the forestry sector until economic liberalization began in the early 1990s were industrial timber management through parastatal corporations and cooperatives, resettlement, and sedentarization of swidden farmers. In the decades after independence, most of the midland and upland area, with a slope over 25 degrees, was allocated for industrial forest management or watershed protection and placed under the authority of state forest companies.

Resettlement policies and programs transferred about 5 million people from lowland provinces into the uplands, in an effort to intensify production on degraded and underutilized forest lands.[161] The Fixed Cultivation and Sedentarization Program resettled 1.9 million shifting cultivators, usually ethnic minorities, in permanent settlements at lower elevation, and encouraged them to adopt sedentary agricultural practices and abandon swidden systems. Mass movements of people, road building, intensifying commercial timber operations, all led to accelerated forest depletion, and by 1991, over 50 percent of Vietnam's state forests lands were classified as barren.[162] The forest policies of this period, emphasizing state and cooperative managed timber extraction and forest land conversion, not only placed immense pressure on remaining natural forests, but undermined the authority and functioning of indigenous community forest management practices.

In many upland areas of Vietnam, "communities" reflect clusters of rural families that may vary in size from a few households to several hundred. These communities or villages are grouped into communes, usually consisting of five to twenty separate hamlets. The commune is the lowest administrative unit in the governance hierarchy and operates under the oversight of the district and provincial government authorities. While communities do not exist as juridical entities, they are often the social loci of indigenous resource management institutions and systems. The absence of formal policies that allow government institutions and programs to easily collaborate with local and traditional resource management organizations has been identified as a limitation in the forestry sector legal framework for over a decade.

As Vietnam embarked on a national strategy of economic liberalization in the early 1990s, the country gradually began to dismantle state forest corporations and cooperatives, allowing private companies and individuals management opportunities through a variety of lease and contract mechanisms that mainly targeted rural households. New policies replacing state forest enterprises with private managers reflected agricultural policy reforms that transferred land management from collectives to private households during the 1980s. The 1993 Land Law gave local families extensive use rights over agricultural and forestry land, stipulating long-term usufruct rights could be issued to such non-state entities as individuals, households, groups of households, and organizations.

A 1994 decree specified that households could have rights to forest land for periods up to 50 years, provided they were used for agroforestry. Farm households were also able to secure long-term management authority for barren lands and forest plantations outside critical watersheds. These policies and programs proved very effective in accelerating the reforestation of barren hillsides in the midlands of the North.[163]

While the 1990s policies and programs that delegated forest management duties to rural households proved effective in supporting plantation establishment and agroforestry in the midlands and some other parts of the country, these policies were less successful in engaging communities in more remote natural forest regions of the uplands.[164] In 1995, Vietnam passed a decree empowering forest administrators to contract farmers and past employees of state forest companies to guard forest lands, a program that later proved to be too costly to continue, and of limited effectiveness.

Strikingly absent from the many forest policies of the 1990s, and from the larger sector reform process, was the direct recognition or support for community forest management. As noted by one analyst from the Ministry of Agriculture and Rural Development:

After the promulgation of the Laws on the Protection and Development of Forests (1991), the State has issued more than 150 important legislative documents with the view to creating a legal framework and encouraging people to actively participate in management, protection and development of forests. The main targets of those regulations are organizations, households and individuals that are involved in the forestry sector. However, as "community" was not legalized as a juridical entity by the Civil Laws (1995), it was not a target to be regulated by those legislations.[165]

Due to their lack of a clear legal "identity," government policies and programs that are oriented towards creating and supporting joint forest management partnerships, find it difficult to develop contractual agreements with communities. In part, formal recognition of communities as legal entities has been slowed by lengthy debate over how they are defined. While communities lack clear legal status, they gained some recognition under the many land and forest related policies approved in the past decade (Table 10).

The absence of legal identity for communities has constrained their ability to play a larger formal role within the emerging national forest management

TABLE 10: Vietnamese Policies and Laws Supporting Community Forest Management [166]

Policy or Law	Strengths	Limitations
Implementation of Law on Forest Protection and Development – Council of Ministers guidelines (Article 9, Decree 17/HDBT-17/1/92)	State recognizes villages as owners of forests, if they belonged to the village for multiple generations.	Only includes forest land that was allocated to the owners by decision of competent public authorities. Most village forests were not allocated by government
Prime Ministerial Decree Decision 327/CT – 15/09/92	Targets communes and villages as the ultimate administrative unit for reforestation and forest management projects at the grassroots level	Village chief used only as a contact point between state organizations and households. Village gains no power or benefits from implementing 327
Civil Law 1995	Recognizes organizations as legal entities if they can meet four criteria: 1) recognized by competent state authority, 2) proper organizational structure, 3) possesses assets, and 4) has independent legal relationships	Does not recognize the village community as an entity with legal status. Neither does the Law on Forest Protection and Development (1991) nor the Land Law (1998)
Forest land Allocation – Decree 163-16/11/99	States that Government allocates forest land to organizations, households, and individuals for up to 50 years.	Since few villages are recognized as organizations, communities are not formally identified as participants in forest land allocation programs.
Regulation of Democracy at the Village Level – Decree 29/CP –11/05/98	Acknowledges that villages "are the living area of local people, and is where direct democracy is implemented to solve internal matters of the community."	States that "A village/hamlet is not a public administrative unit in government structure."
Implementation of Afforestation of 5 Million Hectares – Prime Ministerial Decision 245/TTg and 661/TTg- 29/7/98	Allows contracts with individuals and households for forest regeneration and protection, with benefit sharing of firewood, thinning, and NTFP	Fails to authorize participation of communities or provide guidelines on circulation of products from community forests.
Guidelines for village/community regulations on forest protection and development –	Demonstrates recognition in MARD of the need to facilitate community partnerships in forest management	Attempts to provide guidelines for communities, but fails to extend any legal authority to communities.
MARD Circular 56/TT-30/3/99 and 52/CT-7/7/01 Guidelines for benefits sharing for protection and production forests – Prime Ministerial Decision 178-12/11/2001, Articles 5,7, and 12	Gives specific guidelines for forest product sharing among household and individuals managing state lands, allowing harvesting to proceed.	The policy refers only to individuals or households. There is no clause covering the case of a village community managing a forest.

policies and programs evolving in Vietnam in the 1990s. The constraint reflects not only a lack of support in the Civil Law, but also in the Law on Public Administration, Land Law, and Law on Forest Protection and Development. While the legal and policy framework for supporting community forest management is not yet in place, many provincial and district governments have attempted to respond to the reality that communities have played important forest management roles in the past, are continuing to function as forest stewards, and will be needed in the future in that role. As a consequence, as discussed below, there are cases where local governments are developing management agreements with villages to provide them with greater formal recognition. While local agreements are being crafted to respond to site specific needs for consensus regarding natural resource management responsibilities between local inhabitants and government agencies, the lack of clear policy direction from the central government regarding guidelines for such relationships has created confusion and barriers to the development of a national community forestry strategy.

In 1998, a National Working Group on Community Forest Management was organized in Hanoi, bringing together government policy makers, forestry researchers, NGOs, forest project staff, and representatives of international agencies, to explore how community forest management could be supported through national policies and programs.

Adaptive Implementation of Forest Policies

As is apparent in Table 10, recent Vietnamese laws and policies affecting community forest management provide both advantages and disadvantages for village-based resource stewardship. While the policy movement towards greater privatization during the 1990s has generated new opportunities for rural households to establish a variety of forest management partnerships under government programs, the continuing absence of a clear legal identity for communities has slowed their recognition as formal forest management institutions. Nonetheless, provincial and district administrations have attempted to adapt the policies to provide a greater degree of formal recognition to community groups. A study from three provinces by Vu Hoai Minh and Hans Warfvinge produced under an EC-supported Asia Forest Network project has identified a number of examples regarding such local government-community agreements.[167]

In Hoa Binh Province, for example, provincial officials were eager to implement Decree 02-94 allocating forest lands so that all state forests had an "owner." In Nhuoi village in Da Phu commune, the 125 Muong households had been protecting a thirty-hectare watershed forest for decades. While the Decree did not cover such community forests, the commune People's Committee entered into a formal agreement, delegating management responsibilities to Nhuoi Village. No land tenure certificates with their associated rights were issued to the community. The community, however, did draft a village development plan with a section on forest protection, management, and utilization. A set of forest management rules were agreed on by the community after numerous meetings, and were sent to the commune People's Committee for review, which forwarded them to the District People's Committee for final approval.[168] This case illustrates an attempt by the local government to respond to the center's policy directive to allocate all forest lands to local non-government entities. Finding existing *de facto* management systems already in place, with a need to formally assign management rights, the obvious action for local government to take was to give formal recognition to the Muong community forest.

Khe Ngau village is a community of Thai ethnic minority families located in Xa Luong commune in the uplands of Nghe An Province, in north central Vietnam. The village has 103 households that had been dependent on shifting cultivation and forest product gathering for generations. In 1995, the district Forest Inspection Station initiated a pilot project with the village community to protect 276 hectares of degraded natural forests. The community agreed to protect 120 hectares of evergreen, broad-leaf forest to assist natural regeneration. Yet, to establish legal authority, they needed to find some way to legally secure the land. According to a report on the activity:

> *Given that the formal rules make no provision for management of a natural forest by a village community, the villagers and the staff of the Forest Inspection Station had to find an innovative way to reach their objective: To establish common management of 446 hectares of relatively good natural protection forest in the village. They decided to ask three trusted households in the village to "lend*

their names" by applying for allocation of a third of the forest each. As the entire village had joined the scheme, the request from the three families was not contested by the other villagers and it was also accepted by the authorities.[169]

While the solution was not ideal, in terms of protecting the interests of the entire village, it was an effective way to formalize the forest management agreement between the community and the district government forestry agency.

A third case of local government recognition of community forest management comes from Thy Yen Thoung village in Thua Thien-Hue Province in Central Vietnam. The community borders 2000 hectares of forest along the edge of Bach Ma National Park, which the 400 Kinh families used for fuel wood collection and other non-timber forest products. The villagers had approached local forest inspectors for assistance in protecting the area, as it was under growing pressure. At the same time, the District People's Committee and the Forest Inspection Station were interested in stabilizing forest cover in the area. Since there was no legal framework to allocate forest management rights to the community, the district officers hoped that the provincial government would recognize an agreement on a pilot project basis.

This required a special decision by the provincial People's Committee, which was agreed to on a provisional basis at the end of 2002. The agreement required that the community draft a set of forest management rules that had to be approved by a general meeting of the village, endorsed by the commune People's Committee, and approved by the district People's Committee. The agreement allowed the community to fell up to 50 m^3 of timber annually for production of tools, coffins, house repair, and other domestic needs. The village also held the right to collect dead wood for fuel, harvest non-timber forest products, and hunt animals that were not endangered. A timber monitoring system was also put in place. This model has already been replicated in two additional communities in the district.

This analysis of three provincial experiences in formalizing community forest management systems demonstrates the following points, according to the authors of the study:[170]

- Management responsibilities can be formally delegated to communities.
- Local Governments can legally authorize forest management duties on state lands to communities.
- Communities are sustainably managing forests that local governments have endorsed.
- Agreements and processes that formalize community forest management take different forms, depending on local conditions.

Vu Hoai Minh and Hans Warfvinge concluded their analysis of forest policy implementation in Vietnam with the following statement:

When discussing policies for forestry in Vietnam, it is important to distinguish between policy-writing and policy making. Most policies for the forestry sector in Vietnam are written in Hanoi by senior staff in the Ministry of Agriculture and Rural Development. When they are to be implemented in the field, it is often found that, while they may be good in a general sense, they are inappropriate in specific situations in many localities.

The Government acknowledges this "gap" between the Centre and the localities and, as a consequence, the central government strongly encourages local authorities to adapt policies as necessary to make them work in the field. Law-breaking is not sanctioned, but certainly some constructive policy-bending is allowed...After local authorities in a number of places have developed better policies, effectively making or refining policy, the improvements are incorporated in the next version of the policy by the policy-writers in the Centre.[171]

The question of why explicit community forest management policies have been slow to emerge at the central government level is an intriguing one. It is clear that Vietnam possesses thousands of communities that have traditional or emerging systems of forest stewardship. A recent study from Cao Bang Province in northern Vietnam illustrates how villages in Quyen District have been able to gain support from district governments to restore degraded forests on their limestone mountains (Box 5). While local governments are finding ways to integrate communities into provincial forest resource management strategies, it would logically help them if there was a clear endorsement from Hanoi.

It is apparent that the central government has recognized the limitations of state forest enterprises and

Box 5: The Return of Limestone Forests in Northeastern Vietnam [172]

This case reveals the ways indigenous Nung An people of Cao Bang Province have worked with their district government to strengthen their forest management systems and regenerate their limestone forests. Cao Bang is a mountainous province in Northeastern Vietnam. The karst landscape is a dramatic mosaic of narrow valleys and terraced rice fields interspersed with limestone hillocks that are covered with unique forest ecosystems. Local people value the limestone forests both for their beauty and as a source of water supply for their rice lands. The forests play a critical role in slowing the rate of run-off on the steep slopes, holding moisture well into the dry season. After the limestone forests were degraded in the 1940s and '50s, the hydrology changed threatening local agricultural systems. In the1960s, Nung An communities in the area mobilized to restore the forest, with encouragement from President Ho Chi Minh, perhaps representing one of the first attempts at formal community forest management in Vietnam.

The Nung An migrated into the watersheds of Cao Bang province two hundred years ago from southern China when the region was densely forested. For example, in the 1920s in Phuc Sen village, the limestone mountains were covered in dense forests. As Van Chan, a village leader noted:

> *When I was a child, the elders used to tell me that the commune was covered with jungle. Foxes caught their chickens, tigers caught their pigs, and wild boars trampled on their cassava fields. The forest was public property and can be used by all except the sacred forest where our ancestors lived.*

With low population density combined with isolation, there was minimal pressure on the forest; however, this began to change with the construction of the national highway that opened Cao Bang to the outside world in 1932. Logging pressure, inmigration, agricultural expansion and other forces gradually led to forest clearing and by the 1950s, little remained of the once vast forest area. As one elder recollects:

> *There was no forest, for firewood. We had to go 10-15 kilometers to collect firewood and get permission from another commune ... after some time they began turning us away. It was a very shameful experience.*

Limestone Mountains. Vietnam has 1.15 million hectares of karst, holding 396,200 hectares of the country's remaining natural forests.

> *During Tet festival our town did not worry about meat and rice; we worried about firewood. We had to dig all the tree roots so we could have firewood to celebrate the festival.*

In 1960, the villagers of Phuc Sen began to address the fuel wood shortage after Ho Chi Minh made a call for tree planting. The People's Committee assigned land to villages and households for management. Each village was given two to four limestone hillocks or one side of a mountain to protect and regenerate, while families were assigned degraded land for reforestation near their farms. The community met regularly to plan planting activities relying on several native species, however, natural

MAP 5 : Land Cover and Forest Use Rights, Phuc Sen Commune, Cao Bang, Vietnam

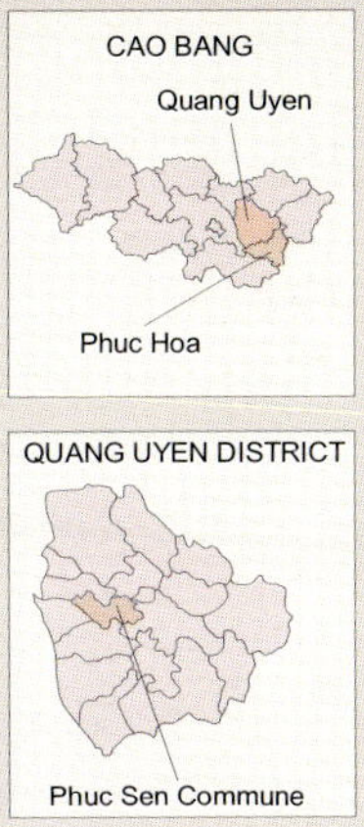

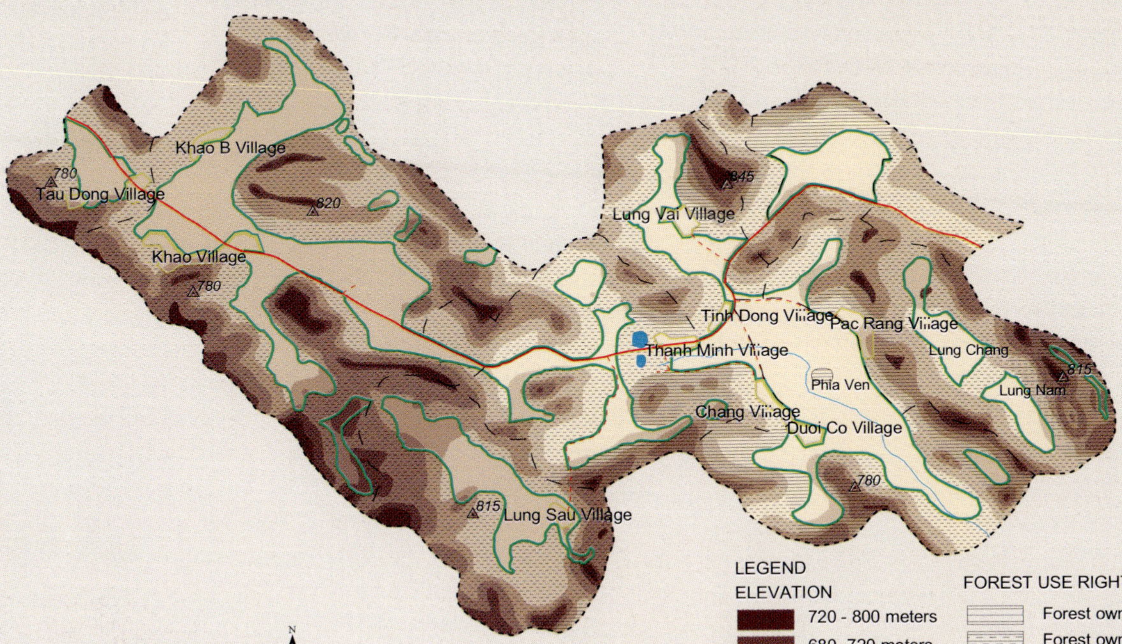

N

0 — 2
Kilometer

UNIVERSAL TRANSVERSE MERCATOR PROJECTION
ZONE 48

LEGEND
ELEVATION
- 720 - 800 meters
- 680 -720 meters
- 640 - 680 meters
- 600 - 640 meters
- 540 - 600 meters
- 480 - 540 meters

LAND USE
- Agricultural land
- Residence

FOREST USE RIGHTS
- Forest owned by village
- Forest owned by household group
- Forest owned by household
- Commune boundary
- Village boundary
- Highway
- Path
- River
- Pond
- 780 Trig point (meters)

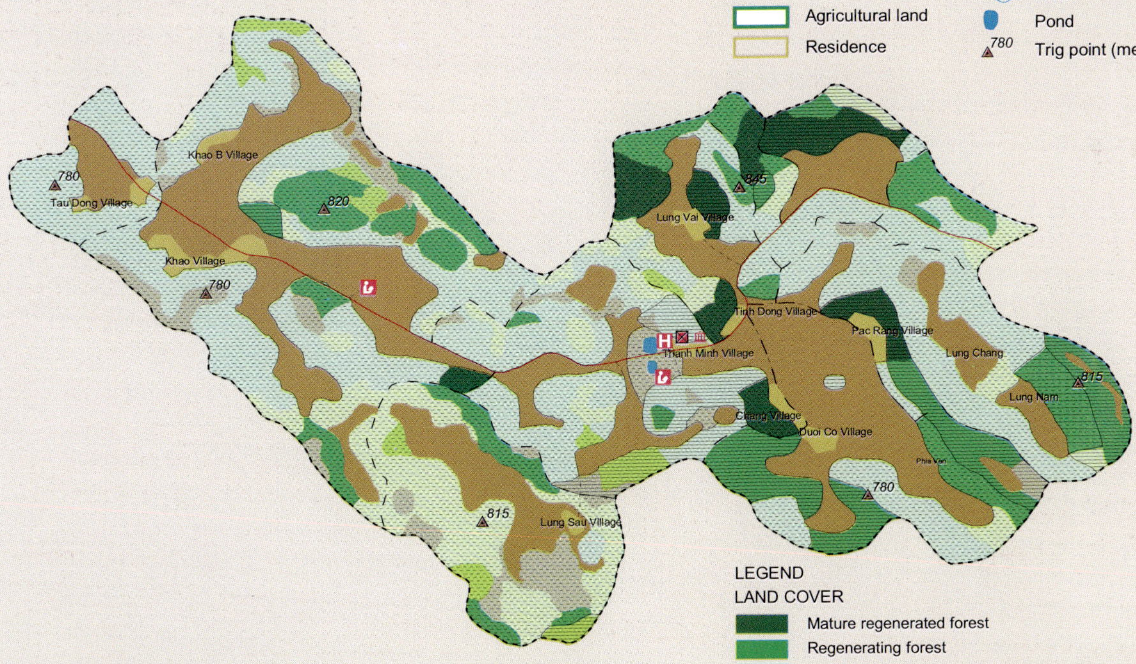

LEGEND
LAND COVER
- Mature regenerated forest
- Regenerating forest
- Regenerating forest with planted saplings
- Bare land with scattered trees
- Bare land with shrubs
- Bare land
- Agricultural land
- Residence

FOREST USE RIGHTS
- Forest owned by village
- Forest owned by househ
- Forest owned by househ

- Commune boundary
- Village boundary
- High way
- Path
- Pond
- 780 Trig point (meters)
- Health station
- School
- Commune P.C. office
- Post office

SOURCES:

Viet Nam Administrative Atlas
BAN DO Cartographic Publishing House

Forest Resources and Environment Center
of the Forest Inventory and Planning Institute
July 2002

PREPARED BY:

AFN ESSC

regeneration also took place and after ten years, much of the area was regaining forest cover. During this period, each village developed rules and regulations to guide forest use. Today over 200 species are found in some areas. Part of the reforestation process involved the planting of trees and natural regeneration of swidden lands. As shifting cultivation gradually declined, so did the incidence of forest fires. Elevation and distance from village in the valley have influenced the type of local ownership agreements, this has also influenced the pattern of regeneration and the spread of the successful management to neighboring communities (Map 5).

By 1990, Phuc Sen was self-sufficient in fuel wood production, while rice and corn yields increased due to additional soil moisture. In 2001, the Forest Resources and Environment Center (FREC) began meeting with Phuc Sen Commune leaders to bring villagers from other communes to discuss their forest management system. The broader goal of the program was to develop a Community Forest Management Network in Quang Uyen District, drawing on Phuc Sen's experience as a source of information and motivation. The first network meeting was held in January 2002. Through this and subsequent meetings, the importance of village-based forest management was revealed. According to the Vice Chairman of Doai Khon Commune:

> *Issues being discussed in this meeting concern us very much. The majority of our commune area is forest land. We share a border with one district and one other commune. Since 1985, we managed 992 hectares of forest, with management responsibilities divided between 4 villages.*

Village forestry and the District network have been welcomed by local government officials and have gained active party support. The series of meetings was effective in articulating the importance of a flexible approach to community forest management. As the District Chief for Agriculture and rural Development (DARD) noted:

> *Rules need not be the same for each village. Rules should be flexible and based on culture, people and characteristics of the ethnic group... They should not come from DARD or the District: they should come from farmer discussions. Meetings should be held regularly to review activities and revise rules so that they can adapt to evolving situations.*

Recently, villagers harvested timber in the village forests to purchase wire to connect their households to the electrical power grid. Now, they are looking for ways to trade with the neighboring provincial markets given the demand for seeds they gather from their regenerating limestone forests. However, the unclear status of ownership in village-managed areas poses insecurity to their current operations. Even though there is a clear mechanism for household-managed forests, they have repeatedly expressed that it is best for everybody if particular forests remain village-managed. While the village forests of Quang Uyen District now enjoy the support of local government and People's Committees, and have aroused the interest of the Cao Bang Provincial Government, these initiatives still fall outside national programs and strategies. For example, the reforestation achievements of the Nung An of Phuc Sen and neighboring communes are not counted as contributing to the national program to "re-green" 5 million hectares. As such, while village forestry is becoming part of local government strategies, it has yet to receive national government recognition.

Traditional systems of CF are long established in Vietnam and in some areas continue to function well. The presence of CF policies and programs, however, could contribute to a strategic resurgence of indigenous forest conservation practices, as well as to the emergence of new spontaneous systems in much of Vietnam's uplands. At the present time, there remains ongoing external pressure to abandon traditional systems of management as other stake-holders (government, timber companies, migrants, etc.) enter remote upland areas. There are also many other potential threats to traditional communities including natural events such as drought and fire, or food shortages that may place greater pressure on the forest. Further, traditional communities are not necessarily cohesive and in agreement as to the use of resources. It is important that the government move quickly to promulgate new legislation and formulate new programs that empower villages to formally manage state lands. ■

has encouraged the transition to people's forestry by initiating policies and programs that support individual and household management. State forest enterprises, however, have not accepted this change easily, many resisting the shift. Provincial government leaders may also find advantages in retaining control over forest resources by continuing to collaborate with parastatal institutions. Finally, many government foresters, especially those at senior levels, continue to question the capacity of villagers to perform roles that had previously been the domain of "professional foresters," a reaction not uncommon in Asian forestry agencies.

National CF Program Implementation

Community forestry in Vietnam takes distinctive forms, just as it does in each Southeast Asian country, reflecting indigenous cultures, as well as political systems and governance structures. The government nationalized large areas in the midland and upland regions of the north in the late 1950s and early 1960s and much of it was allocated to a system of state forest enterprises. State forest enterprises were not structured nor did they have any incentive to work with local communities and governments, preventing cooperation in management between local interests and the forest administration. This created many conflicts and poor management, with the Ministry of Forestry estimating that about 22 million people lived inside or close to the forest in 1986.[173]

Early upland programs initiated in the late-1960s focused on "sedentarization" that involved resettling ethnic minority swidden farming communities into permanent settlements to encourage them to adopt fixed cultivation. Three million shifting cultivators were targeted for the program. National planners, who were usually from the lowlands, viewed upland long rotation agriculture as an inefficient and destructive land use system. As in other parts of Southeast Asia, aggressive government programs resettled ethnic minorities in consolidated communities along roads both to allow them better access to schools and health facilities, as well as to monitor their activities. Communities often benefited from improved market access, educational opportunities, and health care, but found their natural resource activities badly disrupted. The expansion of dams has also caused the displacement of upland communities, as well as inundating forest lands.

By the late 1980s, Vietnam's forestry sector was facing a serious crisis with 10 million of the 19 million of official forest lands classified as "barren."[174] In 1991, the government banned the export of unprocessed logs. Other responses included the expansion of national tree planting campaigns and restrictions on shifting cultivation in many upland areas. Another important approach was the reduction of state forest enterprises and the transfer of greater forest management responsibilities to local government. The dismantling of many state forestry companies reflected the national economic liberalization trend ("*Doi Moi*" or Renovation) that began in the early 1980s. In 2000, two decades after *Doi Moi* started, forest land formerly managed by 412 state forest enterprises in 1990s decreased from 4 million hectares [175] to 3.5 million hectares. State forest enterprises still hold the largest area of allocated lands, however, households, management boards, and other organizations have emerged to collectively manage around 4.5 million hectares of forested land. [176] The State has allocated management responsibilities for almost 8 million hectares, or 72 percent of the nation's 11 million hectares of forested lands. Under this allocation system, management responsibilities of rural villagers are recognized through different means.

The Ministry of Agriculture and Rural Development (MARD) is still continuing to explore community forestry legislative and policy needs, supported by a working group established in 1998 to help inform the process. Earlier participatory forestry programs have focused on households, communes, and other organizations, even though many indigenous forms of community resource management have been hamlet-based. As a result, as in Thailand, much of the effort to formally empower communities as forest managers has been through provincial, district, and commune government.

Official Coverage of CF

In 1983, with the economic liberalization reforms beginning, a new national household forestry program was initiated. By 1990, 4.4 million hectares had been transferred to 2,638 communes, 7,442 cooperatives and nearly 500,000 households. Such programs had dramatic impact in the midlands, as they provided new security to families to manage woodlands. Under Decree 02/CP passed in 1994, land with standing forests was allocated to households for a period of 50 years, while barren land was allocated for even longer periods. The forest land allocation program gave priority to local people, especially those practicing swidden cultivation and by 1992, 800,000 households had obtained land use rights for parcels of forest land.[177] The household forestry program was supported by a national rural credit scheme that extended loans to between two and three million households, with preferential interest rates for investments in upland areas and afforestation. By the end of 1991, non-state organizations, largely households, were borrowing US$489 million. International development agencies also directly contributed to the transformation of the forestry sector committing US$232 million through the year 2000.

The dramatic change in policy from state forest enterprise to household forestry is still in the process of implementation. Part of the strategy involves vesting district governments with managing land allocation and program budgets. Forest land allocation programs have had mixed results. In areas with strong market linkages and better government support, such as the midland regions, afforestation of barren areas has been widespread. By contrast, where upland communities lack adequate road and market access, forest cover and productivity have remained little changed.

At the national level, the deforestation trend that characterized the first eighty years of the 20th century in Vietnam has been reversed. Between 1995 and 1999, the amount of planted forests grew by 450,000 hectares, while natural forests increased by over 1,000,000 hectares. Much of this increase can be attributed to the efforts of local communities and village households who have planted degraded lands, protected and regenerated swidden lands, and established forest plantations. While government programs were not uniformly effective, they were able to address the need to provide forest land tenure security to millions of rural families and to extend financing to restore their productivity. What this approach has not yet directly addressed are the policy and program support requirements for community-based management of larger forest tracts in more remote upland regions. Where community forestry is gaining some official recognition in Vietnam, however, it is doing so through district government based on related national policies such as the Implementation of Democracy at the Village Level (Decree 29/CP (1998), the Decree on Forest Allocation and Lease (163/CP-1999), and other Communist Party Resolutions. A review by the National Working Group on Community Forest Management found that 150 districts had recognized village forest management systems covering 2.4 million hectares of forest land.

Informal Coverage of CF

It is estimated that 27 million people are directly dependent on forests for some subsistence or cash income, representing about 33 percent of the population. These include communities in the Mekong Delta with its vast mangrove forests, hundreds of thousands of hectares of which have been regenerated by communities in the past twenty years; the coastal plains between the Mekong and the Red River Delta, that possess extensive hill forests that cover over one-third of the region and produce nearly 35 percent of Vietnam's timber, much of it processed through small-scale mills run by local community industries, and the uplands which are subdivided between the "midlands" up to 200 meters, and the uplands above.

Since much of Vietnam's natural forests are located in upland watersheds in the northern, central, and western parts of the country, it is these areas where community forestry has the most relevance. While hundreds of publicly managed timber companies have operated logging and operations for much of the past 50 years, covering nearly four million hectares (32 percent of the national forest) in the uplands, large areas of forest above 200 meters remain under the informal management of resident people, especially in areas that are remote from administrative centers and roads. The National Working Group on CFM in Vietnam estimated that around 2.4 million hectares are largely under the informal management of ethnic minority groups. This does not include land allocated to households and individuals, but reflects the vast areas in remote watersheds that are under traditional forms of stewardship.

In the Vietnam context, community forest management can be more aptly referred to as *village forest management*. Vietnamese practitioners further describe forest management driven by initiatives on the ground as *spontaneous management*. The National Working Group identifies the following areas as "village community forests" where spontaneous management is happening:

- Village / hamlet forests that have traditionally been used by upland ethnic cultures for religious purposes. Villagers call these areas "forbidden forests," sacred forests, burial forests or watershed forests.

- Forests previously allocated to agricultural cooperatives wherein residents of one village have managed the area for common use after the cooperative's dissolution.
- Forest land (often with natural forests) acknowledged, respected, or allocated by local authorities to a village for long-term management.
- Forest land (often with natural forests) allocated by Commune Peoples' Committees to a village or a household group within a village without decision from Provincial & District Peoples' Committee.

These forms of village management, however, are presently not yet included in national figures of forest land allocation. The major impediment to incorporating village forest management into national figures is the Civil Law of 1995. Under Civil Law, only organizations with clear organizational structures, recognized or established by authorities, and in possession of assets can be allocated forest lands.

Approximately 50 hill tribes inhabit the upland regions of Vietnam, each with their own resource management traditions and practices. Many of these groups have migrated into the region over the past 400 years from southern China, Thailand, and Laos. Forty-six of these ethnic groups utilize a variety of shifting cultivation systems that require extensive forest management. According to one study of the Hmong and Dzao:

> *Each community has its own sphere of territory...apart from the fixed rocky fields privately owned by individual households, the forest, mountains, streams and rivers are the common property of the community...The community prohibited or limited the exploitation of land or forests within its territory by persons from the outside—especially the utilization of virgin land covered by primary forests.*[178]

While the sedentarization programs initially condemned swidden systems as inefficient and a poor use of resources, there has been greater accommodation of indigenous, community-based forest management in recent years. This is especially true in the north, where ethnic minorities have generally retained positive relationships with the government and the Communist Party. In the south, however, due to the involvement of hill tribes in the "American War," traditional institutions and leaders do not retain the same trust as their counterparts in the north, nor do ancestral domain claims receive the same attention from government. In the central highlands, over one million lowland Kinh were resettled and encouraged to clear the forest for farmland in the 1970s and '80s. During the 1990s, extensive forest areas were converted to commercial coffee estates, some of it with foreign investment. These activities have placed pressure on indigenous community forest management systems in the central highlands. Still, throughout upland Vietnam minority communities continue to be actively engaged in forest management.

The momentum for community forestry in Vietnam has increased with the formulation of the Forest Sector Support Program (FSSP) that is designed to facilitate the emergence of national forest management strategies for multiple objectives. These include protection of the environment, improved livelihoods of people in forest areas, and enhanced contribution of forestry to the national economy.[179] The partnership steering committee includes 16 government representatives and 21 international partners, allowing for better coordination of the government and development agencies working in the sector. There is a general consensus among government planners and the donor community that national policies are needed to 1) confirm the legal status of communities (both at the village and commune level, 2) provide communities with land and forests to manage, and 3) assure that villages and communes receive investment capital and investment advantages to manage their forests. Future laws, policies, and programs will determine to how the government will formally integrate them within the national forest sector.

CHAPTER 12

Regional Summary

In examining the experiences of the five Southeast Asian nations in developing laws, policies, and programs supportive of community forests, it is helpful to return to the five key elements identified by the Workshop on Political Theory that were considered to be central to enable community management of resources. To that end, it is possible to evaluate relevant CF legislation and related laws and policies in terms of their capacity to create a formal framework that supports community 1) rights of association, 2) rights of tenure, 3) rights to access and manage, 4) monitoring and enforcement responsibilities, and 5) conflict management mechanisms. Table 11 provides the authors' assessments of the degree to which key rights are extended to communities within each national CF contexts and through existing laws and policies. This review found that across the five Southeast Asian nations, rights of association are generally weak due to the absence of legal recognition of community level groups, particularly indigenous bodies that have historically played a lead role in forest resource and land allocation and management.

Rights of association have generally been granted to new organizations and institutions that follow "modern" organizational paradigms in terms of structure and function. To that end, CF laws, policies, and programs are typically designed to empower new organizations and institutions, rather than strengthen the rights of association of indigenous bodies. As these laws are applied, communities are faced with a need to articulate the goals, structure, and functions of their resource management institutions in new ways in order to meet the government requirements to establish rights of association, including participating in CF projects, obtaining agreements with forest departments, opening bank accounts, and other formal actions. These requirements potentially create positive incentives for communities to "modernize" their institutions and build capacity, through establishing more transparent processes, securing greater gender and social equity, and developing bookkeeping and accounting functions. The limitations of rights of association so defined are that they tend to bypass positive elements of indigenous organizations and systems, and run the risk of creating parallel systems of resource stewardship that may not be culturally rooted in the community.

Rights of association being endorsed under some new CF laws, however, do recognize the identity and role of sub-administrative village-level hamlets. The Cambodian CF Sub-Decree, for example, defines a community as "a group of residents in one or more villages who share a common social, cultural, traditional and economic interest and use the natural resources in an area" providing considerable latitude to recognize and development management agreements with indigenous hamlets that fall outside the formal governance structure. Ancestral domain legislation in the Philippines also gives considerable recognition to rights and identity of cultural communities, as does the Indonesian constitution, although the latter has had little impact on the ground.

TABLE 11: Forest Rights of Community Forest Groups in Southeast Asia

Key Element	Cambodia	Indonesia	Philippines	Thailand	Vietnam
Rights of Association	Low to Moderate	Low	Moderate	Low	Low
Rights to Tenure	Low	Low to Moderate	Moderate to High	Low	Moderate to High
Rights to Access and Manage	Low	Low	Low to Moderate	Low	Moderate to High
Monitoring and Enforcement Responsibilities	Moderate	Moderate	High	Low to Moderate	Moderate to High
Conflict Management Mechanism	Low	Low	Moderate	Moderate	Moderate to High

Rights to tenure granted under CF and laws, governance legislation, and other legal actions vary among the countries studied. Some of the strongest rights are emerging in the Philippines under the ancestral lands and domain recognized under the revised constitution. Ancestral domain and communal title is also recognized under the Indonesian constitution and the Cambodian Land Law, though only in the Philippines has it been widely implemented. Legislation affecting ancestral domain claims is arguably the most empowering right of tenure, since it may be established in perpetuity. By contrast, most community forestry rights to tenure are time-bound, typically varying from 5 to 30 years, but usually renewable. Nonetheless, they are often subject to dissolution by forestry agencies, though there may be recourse to file appeals. Vietnam provides the longest tenure security with 50-year stewardship granted under the forest land allocation program to households and community groups, while in Cambodia the period is only 15 years, though it is renewable. The Philippines, with 25-year stewardship agreements provides the next longest tenure security to CF groups. In Indonesia, farmers involved in *taungya* (*tumpangsari*) activities, one of the earliest forms of partnership forestry were allowed only three years (until the canopy closed) to plant annual crops, but new programs have extended the tenure period to allow for farmer-owned tree crops. Thailand does have arrangements for CF tenure in production forest areas, though no tenure security has been granted to communities residing in the protected watersheds that cover much of the forest lands in the northern part of the country. Cambodia grants CF stewardship agreements for an initial 15 years, subject to renewal or cancellation by the forest administration. It is too early to accurately assess the degree of tenure security communities will experience under recently approved laws and policies covering community forestry. Ultimately, this will be determined by the ability of communities to withstand external pressures on forest resources and challenges to their authority that will emerge in the decade ahead.

Rights to manage forests are generally low throughout the region, if viewed in terms of the authority of community forestry groups to carry out such routine management functions as forest thinning, selective felling, processing and marketing of timber and NTFP. Most CF laws require the approval of quite detailed management plans before any commercial activity can be implemented. Even once the management plan is approved, forestry agency authorities are often required to sign-off on specific activities before these are conducted. Further, royalty payments and other fees are often required, or a sharing arrangement on harvests is imposed by the forest department. Only in Vietnam are communities given greater latitude to engage in commercial activities that have to do with farm forestry, in part reflecting the national economic liberalization policy. This limitation in current CF law throughout the region is highly restrictive in terms of empowering communities with rights to manage forests. While restrictions and controls on management rights are frequently designed to assure that communities do not overexploit their forest resources, over-regulation may so constrain community ability to manage forest productively, that there will be little economic incentive to invest in management.

Throughout the region most CF laws and policies stress the responsibilities of communities to monitor and protect public forests. Recognizing their limited human resources, forestry agencies realize that they require community collaboration in controlling fires, illegal logging, poaching, and encroachment. The most appealing component of community forestry, to many in the forestry establishment, is the creation of protection partnerships. As a consequence, most CF legislation emphasizes community responsibilities as forest stewards. Ironically, communities often receive limited protection rights under new CF laws. A common complaint among CF groups in Asia is that they have no authority to fine individuals who are engaged in illegal activities, but are only allowed to apprehend them and turn them over to local forestry agency authorities.

Virtually all emerging CF laws provide for the establishment of conflict resolution mechanisms. For the most part these involve a review of complaints by local forestry officials that can be appealed to the relevant ministry if needed. The weakness of such arrangements is that they empower one partner (the forestry agency) to resolve the problem, when the conflict maybe with the forestry personnel. In Vietnam and the Philippines there is a greater involvement of local government in conflict resolution, providing a more transparent basis for adjudicating disputes.

In concluding, CF laws and policies are a recent development in Southeast Asia and it is difficult to accurately assess their efficacy and the degree to which they will empower and enable rural populations to actively manage forests that are claimed as public lands. As the brief review above suggests, most laws and regulations clearly limit the rights and authority of the communities. Given that CF legislation is usually drafted by forestry agencies that are cautious in delegating rights and responsibilities, it is not surprising that initial legal mechanisms are tentative in the authority that they extend. Given the repeated revision of CF regulations in India and Nepal over the past twenty years, it is likely that in Southeast Asia CF laws and guidelines will be repeatedly revised in coming decades as field experiences inform policy makers.

Ultimately, to provide communities with adequate rights of association, tenure, and management, the state must commit to shifting substantive authority from the forestry agencies to the communities. In some nations, like New Zealand and Nepal, forest departments are being radically restructured to facilitate this process. In Nepal, this transition requires forestry staff to switch from a policing and licensing role to one of facilitation and extension. According to the Master Plan for the Forestry Sector the goal involves retraining the entire staff of the Ministry of Forests and Soil Conservation for their new role as advisers and extension workers. Such a transition in the forestry sector of each country will take time and considerable political will. Where political and private sector interests benefit by being able to manipulate the forestry establishment for financial gain, such transitions will be resisted.

Table 12 provides forest area statistics for five Southeast Asian countries, including the area where communities have received some government recognition regarding their management rights to public forest domain. With nearly 40 percent of all national forest lands delegated to community groups, the Philippines currently leads Asia in the amount of area and proportion of total forest under participatory management programs. Much of this is based on the recognition of ancestral domain claims in upland watersheds, as well as community-based forest management agreements. Vietnam has also extended recognition to a larger number of rural households and communities through a variety of forest stewardship programs representing over 20 percent of state forest lands. By contrast, statistics indicate that Cambodia, Indonesia, and Thailand have moved slowly to formally acknowledge the authority of community forest managers, with less than one percent of public forests under CF agreements. The variance in achievements on the ground primarily reflects the different emphasis that national CF programs have received and the stage they are at in created a legal framework to move forward with legalization. Aside from a long-term, ongoing effort to develop a national framework to support CBFM, the Philippines has invested over a

TABLE 12: Forest Cover and Community Forest Management Agreements

	Cambodia*	Indonesia**	Philippines***	Thailand+	Vietnam++
Forest Land (Territory under State Authority- '000 hectares)@	10,535	135,868	15,882	30,143	10,915
Forest Cover ('000 hectares)#	9,335	104,896	5,789	14,762	10,915
Forest Land where some form of CF is recognized by Government '000 has)	228 (fish) 100 (forest)	591	5,900	328	2,348
Percentage of Forest Land Recognized as CFM Areas	1%	0.5%	37%	1%	21%
No. of Groups Involved	767 Villages	NA	4,881 CBFM Sites	1,300 Villages	1,203 Communities

Statistics on Forest land and Forest Cover from:

@ "IUCN WG-CIFM Communities and Forest Management in Southeast Asia. IUCN, 1998. p. 35 or recent national statistics (Cambodia, Philippines, Vietnam). Forest land" refers to territory under the authority of the state. These statistics (except Vietnam) reflect jurisdictional definitions of forest, rather than biological. Much of the "forest land" included in national statistics retains only badly degraded forest cover.

#APFC State of Forestry in Asia and Pacific 2003. FAO 2003. p. 66 and Vietnam MARD Statistics 2000. "Forest cover" statistics reflect land with a minimum canopy closure of 10 to 40 percent, depending on official definitions of forest.

CF Statistics from:

*Tola, Prom and McKenney, Bruce. Natural Resource Assessment in Cambodia. CDRI, 2002. pp. 59, 88-89.

**Includes community forest areas and community mangrove forest plantations. Data for Community Forest Area from Directorate General of Reforestation and Land Rehabilitation (MoF) as of 1995. Data for Community Mangrove Plantations from Forestry Statistics Indonesia (MoF) 1995. (MoF website)

***DENR Forestry Statistics 2003. (DENR website)

+Northern Thailand only. Based on figures from Local Control of Land and Forest: Cultural Dimensions of Resource Management in Northern Thailand by Anan Ganjanapan, Chiang Mai University, 2000.

++CFM Area presents areas implementing CFM models but not necessarily recognized by central government. Figures based on MARD Survey 2001 cited from Workshop Proceedings on National Workshop on a Policy Framework to Support CFM in Vietnam, 2001. National Working Group for CFM. p. 9.

billion dollars in participatory natural resource management programs over the past fifteen years, drawing heavily on financing from USAID and the Asian Development Bank. Vietnam has also relied on the donor community to sponsor a variety of community-based reforestation projects including the Five-Million Hectare program.

By contrast, Indonesia and Cambodia have spent much of the past fifteen years working through their timber concession policies and programs. Momentum and resources have not been invested in CF programs on the massive scale that they have in the Philippines and Vietnam. While Thailand has been engaged in some successful pilot project activities, it has lacked a national CF strategy and program, and suffered from the absence of a supportive legal framework and major financing.

Areas informally under CF may include a wide range of ecosystems, including a substantial amount of degraded forests, lowland forests, flood forests, mangrove forests, montane forests, and even community fishery water areas in Cambodia. While less than one percent of Cambodia's public forest land is reported to be formally under CF, it is estimated that over 80 percent of all forest areas could be managed by rural populations. While Indonesia has some remote forest regions where the scattered distribution of communities may limit CF potential, for the most part, Southeast Asian forests possess a sufficient number and distribution of rural settlements to allow CF to be very broadly applied as a primary mode of landscape management. Furthermore, as has been noted earlier, communities remain the de facto managers of much of the region's forest area. To consolidate management, however, new partnerships between communities, forestry agencies and local government need to be fostered, requiring a much larger investment of political will, human, and financial resources than have been forthcoming in recent decades.

In terms of the impact of state-sponsored community forestry, most national community forestry programs, at least in their initial phases, focus on degraded forest restoration through natural regeneration or plantations. In other cases, projects are geared towards community management of small-scale timber operations. Commercially and biologically valuable forests, typically older growth, has generally been retained by the state for industrial use or placed under protected areas systems. This pattern, however, appears to be changing as environmental organizations find community partners can help conserve biodiversity, and as good forests in critical watersheds are taken out of timber production in order to achieve other management goals.

Beyond general figures for these country area "accomplishments," it is difficult to measure the social impact of community management. Community forest management still largely belongs to the informal sector not only in terms of its economic relevance, but also in terms of overall government recognition and national awareness. While anecdotal stories of success and failure abound, no Southeast Asian nation has yet developed an effective national system for monitoring and evaluating the social, economic, and environmental impact of community forestry initiatives. As a consequence, it is difficult to accurately evaluate the broader impact of CF at a national or regional level.

The next section discusses some of the forces that are both facilitating and constraining such longer term transitions in the forest sector, as well as a number of important contexts in Southeast Asia where community forestry could be implemented to stabilize threatened environments.

PART 5

Forest Transitions

National political and economic changes in governance structures, public opinion, and markets appear to be enabling a greater involvement of community in the formal management of state forest lands in many parts of Asia. Part III and Part IV documented the increasing number of laws and policies that either directly enable CFM activities or encourage community resource management through related legislation and programs. While some specialists argue that these actions have not gone far enough in empowering communities,[180] it is apparent that the forest-dependent peoples are considerably stronger from a legal standpoint than they were one to two decades ago. Due to the change in the overall legal and policy environment affecting indigenous peoples and forest dwellers, their tenure security has been enhanced though overall forest cover has been clearly degraded for the complexity of reasons already given.

There is also no doubt that external pressures, both from the private sector as well as governments, has grown steadily over this period. Demand for land, timber and mineral resources is unabated, and while environmental policies may restrict exploitation, new roads facilitate access. Where deforestation has slowed in many areas it is more likely due to the exhaustion of forest cover, rather than improved protection. Yet, as each national story suggests, communities in some areas are slowly gaining better control over forest resources. Whether these experiences will be replicated to form national rural people's movements for better forest management practices or remain isolated exceptions to current trends, remains to be seen. In examining case studies from Southeast Asia presented in Part IV, it is apparent that certain factors are associated with the success of CF in terms of forest restoration and protection, enhanced tenure security, and increased productivity, while other forces constrain transitions in the forestry sector, both at the local and national level.

Part V examines some facilitating and inhibiting factors currently at work in Southeast Asia. The effectiveness of CF implementation may be associated with the presence of traditional systems of resource stewardship and governance, newly emerging leadership and strong local organizations, effective external support, or a combination of all of the above. This section reviews how the presence of such factors creates "hot spots" of community forestry that provide a basis for anticipating where CF policies and programs might be most successful. Reviews on community forest management cannot be based simply on the implementation resulting from the execution of CFM policies, programs and projects but also from related programs and projects that possess overlapping objectives and strategies and target communities in forest areas. As Part II indicated, in many Asian nations devolution of resource management rights to local governments is driven more by decentralization policies and laws, than by specific CF policies and programs.

Law and policy implementation, especially in developing countries, is often governed by the programs and budget allocations that give it operational definition. Most programs do not consider the nature of local initiatives, nor are they fine-tuned to respond to local contexts, but rather present the "generic" national strategy. The range of community contexts largely reflects the historical experiences and social, economic and physical characteristics present in each area. The capacity of communities to effectively relate to opportunities that may be present in CF programs and agreements may vary widely, while the extent to which national programs are a "supportive fit" may also differ across the country. In some regions, CF programs may respond poorly to local problems and needs.

The process of establishing community forest management plans is less complex in social contexts where indigenous systems of resource stewardship are well articulated and retain a high degree of social acceptance. In such cases, the primary need may be to document and codify existing forest tenure arrangements and practices. The acceptance of traditional resource management systems by local government, however, may not provide the security communities ultimately require in the face of competing claims, especially if they come from more powerful, outside political actors. Indigenous management practices and traditions provide a basis upon which to build new CF systems, while new forms of community leadership and strong external support can also catalyze the emergence of effective CF initiatives. CF "hot spots" can emerge in response to different management goals. CF activities often reflect efforts of a community to change the status quo, especially when they are initiated by the community themselves. They may attempt to protect degrading forest lands, enhance productivity, resolve resource conflicts, or strengthen tenure rights. In other words, site specific problems and issues are often motivating factors that drive community participation. The reduction or resolution of these problems often reflects the success of the community's efforts to mobilize their resources to address them.

CHAPTER 13

Supportive Forces

The capacity of communities to manage "state" forest lands is influenced by external forces that constrain or support their efforts. Supportive forces may include development agency projects, program budget availability, NGO facilitation, civil dialogue forums, and research and monitoring.

Civil Society

Weak political will, vested interests, corrupt practices, and limited government capacity often constrain the effective devolution of meaningful forest management rights and responsibilities to rural communities. In response, a growing number of communities in Asia are organizing associations, federations and networks to gain greater political voice. Such civil society institutions typically appear after a substantial number of communities have already experienced some success in obtaining control over natural resources in their area. Community and NGO alliance building through networks and federations is also an increasingly common phenomenon in Asian nations. Communities in Nepal, northern Thailand, eastern India, northern Vietnam, northeastern Cambodia, and the southern Philippines, among many other places, are forming networks and federations in order to develop broader action strategies to address more powerful stakeholders, whether they are forestry agencies, private sector firms, or local government representatives. While a single hamlet of twenty to fifty households may have little chance of challenging the illegal or legal forest use operations of outside stakeholders, if they are part of a larger association of ten, fifty, or one hundred villages or more, they can mobilize more visible protests and are also more likely to attract the support of elected or appointed officials.

Forestry agencies, local government and development organizations have opportunities to extend and strengthen CF management and production systems by catalyzing the formation and development of community resource management networks. Patterns for such association may vary, based on socio-cultural, economic, hydrological, or administrative factors, though they should be designed to establish meaningful social units of communities with shared needs and aspirations. Given decentralization and democratization trends present throughout much of Southeast Asia community-based resource management associations may gather strength through their electoral influence, allowing them to better challenge external vested interests that compete for control of the forest. At the same time, experience from India and Nepal suggests that such associations are vulnerable to manipulation by individual political parties.

The growing number of non-government organizations that have incorporated community resource management as part of their program strategies has increased capacity in many Asian nations, as financial support for CF expands. NGOs are also actively contributing to forestry policy discussions throughout the region on a range of issues. Some international NGOs formerly engaged in relief assistance in rural areas such as Care, Oxfam, World Food Program and Feed the Children restructured program activities to support communities engaged in natural resource management. NGO strategies to strengthen community forest management include policy advocacy, awareness-raising, and institutional capacity building. Now that policy development and tenure security are slowly being given attention, skills in facilitating resource and livelihood planning are needed to bring communities to a further level of stability.

As discussed earlier, many countries in Asia have formed national working groups to serve as policy forums that help inform community forest management transitions. These mechanisms are helping build consensus among key stakeholders in government, university, NGO, and donor agencies.

Indigenous Culture and Local Knowledge

Most rural communities in Southeast Asia have established traditions of natural resource management that are adapting as the political, economic, and natural environment change. In many cases, rules and norms of resource utilization have been evolving for centuries, and that change is accelerating as the rate of social transformation intensifies. Many traditional guidelines for community use of resources attempted to ensure a level of social equity, prevent the erosion of resource control to outsiders, and to allow for sustained levels of production of timber, non-timber forest products, plants, animals, and water. As population pressures have grown, outsiders have entered the area, and commercial markets have expanded, communities have found a need to intensify resource management controls. This often involves the further delineation of rules, tightening of access regulations, and the formal codification of use guidelines. In some cases, the emergence of "by-laws" based on customary or informal agreements between communities, forestry agencies, and local governments have provided a greater degree of security for resident communities. The formalization of community forest management rules by local villagers is a widespread and growing phenomena in many parts of Southeast Asia, especially where forests are endangered. Many indigenous communities that have historically held land collectively are seeking ways to relate customary tenure to modern land markets that emphasize private titling.

Lindsay, however, notes that traditional community management rules and regulations do not provide long-term security to the community as such regulations have no legal authority over outside stakeholders, whether they be other communities, timber concessionaires, or national government agencies. While nations like Indonesia provide constitutional recognition of customary laws (*adat*), community resource claims can be overruled by the national government. Lindsay illustrates these limitations by referring to the language of some national laws that limit the authority of community forest ordinances as follows: "customary rights of forest-dwellers will be respected as much as possible" or "customary law shall be respected unless the national interest requires otherwise."[181]

In the absence of national CFM policies, community efforts to review, update, and codify customary laws may provide them with additional tenure security, especially if alliances are built with neighboring villages and local government. Nonetheless, to obtain meaningful legal security over forest resources, community rights need to be enforceable against claims made by the state and outsiders.[182] It is unlikely that, in any nation in Southeast Asia under existing civil laws, customary forest laws would stand up in formal courts if confronted by competing claims of the state. Even most of the recently approved CFM policies would be unable to protect the rights of communities if contested by higher law.

Market Transitions

Making a forest more productive is not only a desirable goal for low-income forest-dependent peoples, but also helps ensure that forest environments will be sustained in the future. Southeast Asian governments have limited resources to invest in forest management, consequently communities need to find ways to generate revenues from the forest to meet both their own needs, as well as those incurred in managing the resource, including fire protection, thinning operations, enrichment planting and other silvicultural and agroforestry activities. Governments must extend sufficient tenure security and management rights to allow communities to invest in their forests and make timely, market-oriented decisions, while ensuring that any royalties or fees imposed allow communities an attractive profit margin.

Southeast Asia's forests have long been managed productively and for thousands of years non-timber forest products were the source of most of the region's important exports. Over the past five hundred years, indigenous systems of agro-forestry have evolved, many of which combined subsistence goods with export commodities. Mimicking natural forests, these agroforests are often complex in structure and biodiversity. Unfortunately, because they do not reflect the economies of scale inherent in large cash crop plantations or timber concessions, they have not received the support or investment capital that has been poured into the former. Nonetheless, economic analysis has shown they are far more productive per hectare, sustainable, and present far less risk. The lack of external support for indigenous agroforestry in Southeast Asia reflects the unwillingness of donors and private sector investors to collaborate directly with rural communities. In part, their reluctance is due to concern over control issues. Private sector investors want to control the profits, and development agencies and governments want to have control over their investments. Nevertheless, millions of hectares of degraded forests need restoration in Southeast Asia. Agroforestry systems represent a highly appropriate management approach in many areas, responding to livelihood, hydrological, biodiversity requirements. Therefore, capital investment in agroforestry development should receive higher priority from government, private sector, and international donors.

Non-timber forest products continue to have tremendous economic potential in Southeast Asia, but much of that has not been realized due to limited capacity to process and market these goods. Estimated value of the international rattan trade is valued at US$ 2.7 billion a year with most of the production from Southeast Asia.[183] Ecotourism has also grown rapidly in Southeast Asia, and is often viewed as a potential source of livelihood for forest-dependent peoples. Experiences from northern Thailand, where ecotourism has been operating for several decades, are mixed. In Thailand, as in many other parts of the world, the problems are numerous, but two of the most critical ones are related to negative social influence and inequities. Inviting foreigners and urbanites into small, rural villages is often disrupting for the community, while tour operators tend to capture the vast majority of ecotourist expenditures. More successful have been the coastal initiatives involving mangrove areas where communities have more direct control over the terms for visiting groups and where there is less negative cultural impact. New project paradigms need to be developed in Southeast Asia that support community-based forest restoration that leads to productive management and equitable returns.

Development Agency Support

The effectiveness of CF Policy development and implementation often reflects the willingness of development agencies to finance these new initiatives. In the case of community forest management, as discussed in Part III, donors have made forestry sector policy change a pre-condition for new funding in some countries. Donor agencies and multilateral organizations are currently giving more attention to livelihood needs of millions of upland people. International conventions are integrating social concerns in their environment and biodiversity agenda.

National governments in most Asian nations are not yet in a position to provide and sustain direct support to communities engaged in forest management as it is not a program that can generate immediate and tangible revenues. However, governments are increasingly under pressure both to account for resource allocation and to protect areas that require the engagement of local people. Development agencies have played a strategic role in financing CF initiatives in such areas, often through their poverty reduction programs. Development agencies such as the European Commission, World Bank, ADB, USAID, OECD, GTZ, SIDA and DFID exert considerable influence over government development priority setting. Policies supportive of community rights and responsibilities in forest management are often driven by donor agency funding.

Beginning in the early 1980s, some international organizations began supporting community-based resource management strategies in the Philippines, Indonesia, and Thailand in an effort to assist governments to develop more sustainable and cost-effective stewardship systems for state lands, while responding to the needs of the rural poor. At that time, the sustainability of industrial timber extraction practices was just beginning to be questioned, and most development agency funding in the forestry sector still emphasized better practices for commercial logging. There was very little recognition of indigenous systems of forest management, ancestral domain rights, or even collaborative models of forest management.

In the late 1980s, USAID began a large support program for community-based natural resource management that has continued up to the present with an emphasis on Indonesia and the Philippines. Throughout the 1990s, the programs of the Manila-based Asian Development Bank began to incorporate community forest management strategies. More recently, the European Commission, through AIDCO, allocated US$300 million for tropical forests in developing countries for seven years starting 2000.

> *The European Community, along with all major donors, is committed to meeting International Development Targets, which include halving the proportion of people living in absolute poverty by 2015 and reversing current trends in the loss of environmental resources. The sustainable management of forest resources, and activities involving the millions of poor people who depend wholly or partly on forests for a living, will make a contribution to meeting these development targets.*[184]

Recently, the European Union has intensified efforts to review its policies and that of its member countries to combat illegal logging and promote certification systems. Slowly, EU policies on trade and forest certification are having an impact. Top Asian wood exporters are setting up their own certification systems and marketing their credibility in response to new EU procurement policies.[185] Meanwhile, entrepreneurs are also crafting ways to get around the system. Some Vietnamese companies import certified lumber from Malaysia, process it into garden furniture, then distribute it throughout Europe as products coming from certified sources.

The UK Department for International Development (DFID) has adopted a Renewable Natural Resources Research Strategy to help country partners eradicate poverty by supporting research on priority developmental problems of the forest-dependent poor. The Forestry Research Program, one of ten competitive grant portfolios financed under this strategy, aims to improve the livelihoods of poor forest-dependent people in tropical and sub-tropical countries by supporting studies on sustainable livelihoods, tropical and sub-tropical forestry management, watershed management, carbon sequestration, tools for clean development mechanisms, and non-timber forest products. [186]

The World Bank forestry portfolio has evolved away from an emphasis on industrial timber extraction during the last twenty years. In the 1970s, the Bank had a narrow focus on industrial plantations, with 13 of 17 loans devoted to enhancing the timber supplies for emerging markets. In the 1980s, the World Bank earmarked only 40 percent of its $500 million forestry budget for industrial plantations, and concentrated the rest on fuelwood production without much attention to the larger policy context. In the 1990s, however, the budget for timber production was decreased to 20 percent and the remainder allocated for social forestry and environmental protection programs. In another assessment based on IFC data, it was found that between 1991 and 1998, only $578 million dollars were loaned for forest industry investments out of a project cost of $3 billion for the sector.[187] The Forest Sector Strategy of 2002 and the recent World Bank Environment Strategy also identified natural resource management as a priority area for rural development initiatives:

> *Despite more than a decade of attention, policymakers in the East Asia and Pacific region are just beginning to recognize the importance of natural resource management issues for sustainable economic development... While many agricultural*

TABLE 13: World Bank Forest Sector Transitions

	1953-1976 (1)	1978 Forest Sector Strategy (1)	1991 Forest Sector Strategy (2)	2001 Forest Sector Strategy (3)
Strategy	– Projects in Africa and South America – Industrial timber plantations – Rural development – Forest conservation	Bring about a change that will result in a larger share of resources being allocated to rural afforestation programs by influencing policies of developing countries	– Multisectoral approach – International cooperation – Policy reform and institutional strengthening – Resource expansion and intensification – Preservation of intact natural forests	– Harness potential of forests to reduce poverty – Integrate forests in sustainable economic development – Protect vital local and global forest environmental services and values.
Loan Portfolio	$240M (1953-76) 13 of 17 loans towards industrial forestry (pulp & paper)	$500M (1979-83) 40 percent for industrial forestry activities	$3700M (1990-2000) 20 percent for industrial forestry activities $788M to East Asia Pacific	$848 (2001-2005)
Conditions	Loans based largely on cost/benefits and rates of return	Funds not given to logging in tropical moist forest	Environmental Impact Assessments Required	Environmental and Social Impact Assessments required

Sources:
(1)World Bank Forestry Strategy 1978
(2)World Bank Forest Sector Review 1994
(3)The World Bank Group. A Revised Strategy for the World Bank Group. 31 October 2002

development projects, particularly those in China, have been successful in raising farm-level incomes, they have not generally been effective in promoting sustainable NRM. The challenge is to redirect rural development initiatives away from individual agricultural production projects toward sustainable NRM.[188]

Table 13 illustrates the changes in World Bank forestry sector strategies over the past 50 years.

The Asian Development Bank (ADB) has increasingly enunciated the importance of forests for both forest products and environmental services. ADB loaned almost one billion US dollars to Asian government forestry organizations from 1980 to 1996. Over this 16-year period, ADB's forest sector strategy shifted its emphasis from increasing forest productivity to enhancing forest sustainability. The shift of focus is partially related to widespread criticism from NGOs and academics that ADB had spent too much financing large-scale plantations by government and subsidizing private sector logging operations, rather than on smaller, community-based schemes to rehabilitate degraded natural forest.[189] The forthcoming ADB Forest Policy[190] appears to be far more supportive of community forest management. The forestry strategy is required to support the ADB's overarching objective of poverty reduction based according to the 1999 ADB Poverty Reduction Strategy. Under the new strategy, greater programmatic focus will be placed on people living in or near forest areas. Land and forest tenure security is being highlighted as an important governance issue for securing the conservation and development objectives of ADB.

Despite a substantial investment in the forestry sector, there remains little hard evidence demonstrating how past investments contributed to the reduction of poverty, the strengthening of community tenure security, or the improved management of natural forests. In part this problem reflects the relatively recent redirection of forestry sector support to address community needs, and the decades required to reorient the management of hundreds of millions of hectares of forest. From the 1950s through 1980s, and in the case of Cambodia until recently, development agencies have supported the development of commercial timber industries. At the same time, the forces driving deforestation and the displacement of upland, forest-dependent peoples, are in many ways more powerful determinants of behavior, than the time-bound, site-specific projects. Finally, attitudinal shifts within government and donor agencies are still under way, with a significant proportion of professionals continuing to view local communities as a problem rather than a solution for forest management. In such instances, implementing staff undermine the achievement of the people-oriented program goals.

While there are many obstacles to effective development agency support for forest sector reform, over the past two decades there has been a clear shift in donor orientation towards community forestry. This is evidenced by the amount of funds flowing into participatory resource management, the amount of innovation apparent in their design, and the number of professionals with experience and expertise relevant to CF program needs. Provided donors continue to allocate substantial resources to community-based resource management in Asia, it is likely that the impact of such initiatives will expand. Recent donor efforts to coordinate country strategies and link livelihood and rural resource management projects may encourage the expansion of CF groups in some countries, and accelerate the development of small scale forest-based enterprise.

International Commitments and Conventions

The United Nations Forum on Forests (UNFF), created in October 2000 after ten years of discussions among governments, offers an intergovernmental platform for policy dialogue and promotes the implementation of more than 300 proposals for actions of its predecessor, the Intergovernmental Panel on Forests/ Intergovernmental Forum on Forests (IPF/IFF). Of the 300 proposals for actions, there are several points that can create a positive impact on CF if they are adequately clarified and given focus. Some of the actions that express a level of support for CF:

- Address the underlying causes of deforestation and forest degradation within each country, including the impact of poverty and the impact of processes outside the forest sector;
- Create appropriate procedures in order to promote effective participation of all interested parties in decision-making about forest management;
- Support appropriate land tenure law and/or

arrangements as a means to define clearly land ownership, as well as the rights of indigenous and local communities and forest owners, for the sustainable use of forest resources, taking into account the sovereign right of each country and its legal framework;

- Develop mechanisms, as appropriate, to improve land access and use of forest resources on a sustainable basis;
- Support capacity-building in communities, in particular for those with responsibilities in forest management, including low forest cover countries, and create awareness in the society at large on the importance of issues related to deforestation and forest degradation;
- Support and promote community involvement in sustainable forest management through technical guidance, economic incentives and, where appropriate, legal frameworks;
- Support local community programmes for capacity-building and credit facilities, and facilitate access to domestic and external markets of forest products and services.[191]

The process and negotiations emerging under the United Nations Forum on Climate Change, including the Kyoto Protocol and the Clean Development Mechanism, have attracted growing attention to deforestation problems as a source of greenhouse gas emissions, as well as to the potential of afforestation and reforestation programs to enhance the effectiveness of terrestrial carbon sinks. Under the Clean Development Mechanism, there is now a growing possibility that rural households and villages might be paid to re-establish forests, offering a new source for financing community forestry activities. While the prospects of establishing a global mechanism for environmental service payments to the upland poor are attractive, many barriers are present and the uniqueness of each situation requires particular responses.

The Convention on Biological Diversity commits not only to biodiversity conservation and sustainable use, but also to equitable benefit sharing. Goal 4 in the Forest Work Program approved by the 6th Party to the Convention on Biological Diversity reads: "Enable indigenous and local communities to develop and implement adaptive community management systems to conserve and sustainably use forest biological diversity."[192]

The Global Caucus on Community-based Forest Management emerged from the lobbying efforts of a number of advocacy groups participating in the World Summit on Sustainable Development (WSSD). The Caucus aims to build a coalition of organizations that can inform international forestry agreements, as well as global strategies and mechanisms to better support community engagement in forest management. The Caucus also intends to establish links with the mass media to help inform public opinion and create awareness regarding the potential of CFM.[193]

International agreements, global programs, and advocacy movements are all shaping the national contexts in which CFM policies are formed and projects implemented. Recent international forestry and environmental agreements ratified by the United Nations clearly articulate the rights of indigenous peoples and forest-dependent communities to forest resources and encourage their formal engagement in stewardship, both to facilitate sustainable management and to generate livelihood. The UN Millennium Development Goals and the World Bank Poverty Reduction Strategy Papers are also lending support to community forest management. These goals, at least on a rhetorical level, are being gradually integrated into policy documents in a number of Southeast Asian nations, encouraged by development agencies. International agreements have been less effective in generating additional financing for CFM, as witnessed by the outcome of UNFF negotiations. Nonetheless, the formal agreement of the international communities to support transitions to greater community involvement in forest management is having an impact on Southeast Asian policies, if a lesser influence on their implementation and effect on the ground.

CHAPTER 14

Constraining Forces

While a number of forces described in the preceding pages are supporting an Asia-wide transition towards community-based management of state forest lands, many factors continue to constrain this sector shift in management orientation.

Political Will and Corruption

"Political will" in this context refers to the commitment of the political leadership to implement policies and laws that have been adopted by the nation. This political will may be exercised at the central government level, or it can be reflected in the actions of local government officials. In many cases, laws and policies are not implemented due to the absence of political will, especially if they were adopted during election periods to bolster populist campaign agendas or forced upon national leadership by development agencies. This is a significant problem with the leveraging of forest sector policy reforms through structural adjustment loans, whether promoted by the World Bank, the International Monetary Fund, or the Asian Development Bank. To a lesser extent, micro-policy reforms have been leveraged by bilateral agencies by tying such actions to the approval of a major forestry sector project. An absence of political will may manifest in many ways including a reluctance to disseminate information widely regarding new policies and laws, authorize funds for field programs, allocate staff for CF activities, set targets for local officials, monitor programs, or cancel logging concessions and create opportunities for CF expansion.

Efforts of development agencies to encourage CF by facilitating forestry sector reform by offering structural adjustment loans and projects have met with mixed results. Despite economic incentives provided through such loans, vested interests may resist and undermine implementation of national CF policies for years, or capture them to meet their own ends. In many Southeast Asian nations, particularly after independence was achieved, forest resources were one of the major sources of revenue for governments as well as their leaders. As discussed in Part III, national political coalitions were often forged with the allocation of forest and land concessions acting as a major bargaining tool. State forest land continues to be a major commodity for political transactions and coalition building, whether it is allocated for timber, non-timber forest products, access to mineral resources, or land for the establishment of industrial estate crops.

Implementation of forest policies that threaten current resource access relationships are often rejected by individuals and institutions with interests in existing private sector operations. Since forest resource access has often been controlled by alliances of highly placed government officials, forest agency staff, and private sector stakeholders, such groups can effectively stall the implementation of CF policies and laws. As a consequence, even though industrial forest management modalities have proven to fail in terms of profitability or effectiveness, they still deliver substantial income to well-placed individuals who may defend them vigorously.

In the Philippines and Indonesia, both former Presidents Marcos and Soeharto enriched their families and built political and economic alliances through the allocation of logging rights. Corrupt alliances involving state leadership and the private sector have been common practice in many Southeast Asian countries for decades, especially in the forestry sector. Forests became increasingly susceptible to corruption by government officials as these lands were nationalized from the mid-19th century onwards. By overriding community rights through a process of state-imposed demarcation and reservation, state forest agencies were established as local managers ultimately under the control of the political leadership. As a consequence, political leaders could reallocate use rights to "friends" in control of private enterprises. This relationship has been referred to as the "iron triangle."

Throughout Asia, parastatal forestry companies persist, despite dismal performance in achieving their own state objectives including: 1) generating revenues for the state, 2) sustainably managing natural forests and plantations, and 3) generating employment and responding to the needs of rural communities. These institutions have, however, been effective in meeting their prime, unspoken goal of channeling forest-related revenues to elites in the government, private sector and military. Since parastatal organizations are relatively easy for those in power to manipulate, they are effective institutional tools to capture forest resources. At the same time, the continuing presence of state forest organizations in the forestry sector has been an important impediment to the effective implementation of community forest management policies and programs. Perum Perhutani's (Indonesia's State Forest Corporation) successful lobbying efforts that led to the Ministry of Interior's decision to nullify the Wonosobo District Government's legislation and program to reforest degraded state lands, is a prime example of how a parastatal organization may undermine highly successful community watershed restoration, primarily because it threatens their control.

In Thailand, for example, the government-owned Forest Industry Organization (FIO) has long been criticized due to widespread corruption in the organization.[194] Founded in 1947, this state-owned corporation generated revenues by managing timber concessions in natural forests. Under the FIO and private operators, Thailand's forests were rapidly depleted during the 1970s and '80s. In 1990, the government imposed a logging ban forcing the company to rely on plantation management, and it has been losing money steadily since that time. According to a recent report, "The organization is shouldering more than one billion Baht in debt ($25 million). FIO is unlikely to get rid of the debt if corruption and malfeasance are widespread."[195] The report goes on to note that "...suppression of corruption in the FIO is difficult because the activities are backed by senior people on the executive board."[196]

Clearly, the devolution of forest management rights and responsibilities back to local communities is inherently threatening this alliance, as it shifts control away from the political leadership back to rural villages. CF policies, if implemented, result in the creation of local management plans and structures, where forestry agencies are no longer responsible for day-to-day management decisions. If these policies are properly implemented they lead to the formulation of community forest management plans which determine what activities can and can not be carried out in a given forest area. All of these actions directly undermine the local control and authority of field level forestry staff, and ultimately the ability of political leaders to transfer exploitation rights to cronies in the private sector.

Decentralization whereby resource planning and management authority is transferred to local government may not solve the problem of lack of transparency and corruption. Provincial governors, district chiefs, and even village heads may face similar needs to buy support with the resources, or pad their personal coffers. Forest resources in a province or district are in most cases seen as a source of revenue for local government and the number of checkpoints increases along with the level of illegal fees on any forest or plantation resources. For these reasons, it is evident that political leaders, at whatever level, who have found personal enrichment or power through authority over state forest resources, or who support their position of power through their influence over forest resource allocations, might have less incentive or "political will" to implement CF policies.

Bureaucratic Barriers

While CF-related legislation provides a legal framework for formal devolution of resource stewardship rights and responsibilities, management plans and formal agreements make devolution a reality. The approval of an integrated management plan is often viewed as the culmination of the planning process, one that is distinctly different from conventional, silviculture-oriented planning which focuses largely on evaluating growing stock through forest inventories. CF plans required by new government laws often involve a wider range of forest services including NTFPs, fuelwood, hydrological function, and even biodiversity management. Multiple methodologies are used including participatory assessment activities to identify management priorities and needs, community mapping to fix territorial boundaries, and the formulation of user rights and rules. The level of precision required in the preparation of CF plans varies between countries and is still under development in many nations, but frequently far exceed local capacities. Experience with CF management planning has been mixed in Asia.

In the Philippines, Claims of Ancestral Domain Certificates (CADC), and Community Forest Management Agreements (CFMA) confer limited rights on participating groups until management plans have been formulated which comply with the complicated guidelines set down by the DENR. Philippine CFM planning procedures are sufficiently complex that most communities find it difficult to compile a plan without the help of an NGO or university group experienced with the process. DENR officials under pressure to generate community plans for project funding are driven to draw up *pro forma* plans themselves. The fast- tracked accomplishments result in plans that communities have not internalized and hence can not implement in any meaningful way. While each plan is reviewed by a technical committee in the DENR, even if it meets the technical requirements implementation may be delayed indefinitely for political reasons or suffer the fate of total extraction bans.

In India, micro-plan processes were established as part of the Joint Forest Management program in the early 1990s in order to delineate an agreed set of management practices for communities and forest departments. The micro-plans are intended to represent an operational management agreement for implementing JFM guidelines, not to mention a fundamental element in many donor programs, including those of the World Bank. Thousands of micro-plans were formulated in India over the past decade, often with disappointing results. The influx of large forestry loans from development agencies put pressure on state forest departments to rapidly formulate the plans in order to release project funds for capital expenditures on plantation establishment, silvicultural treatments, watershed structures, and related investments. As a result, rather than emphasizing a dialogue among community members regarding resource management priorities and long-term forest stewardship goals, the planning process was directed towards the identification of short-term requirements for project implementation. Further, the process of "micro-planning" was typically directed by, if not completely implemented by, forest department field staff. This reduced opportunities for community discussions, capacity building, and joint formulation of long-term management goals. In addition, the management goals were often set down by the state forest department according to often archaic Working Plans, which have guided silvicultural management in India for over a century, but maybe increasingly outdated in India's post-industrial timber era.

In India, it is less a problem that the methods require technical data beyond the capacity of communities, but more that forestry field staff usually have little experience or ability to facilitate community management discussions. They are far more comfortable working alone on conventional planning tasks, especially when under time pressure. To that extent, expecting Indian forestry field staff to quickly switch from a technical and policing role, and instead to act as community development agents, has not been realistic. Tanaka points out that community forest management plans are needlessly technical, reflecting traditional biases in forestry agencies for timber management planning.

The prevailing problem is that the types and depth of information required in the management plans are often unilaterally pre-determined under the

TABLE 14: Community Forest Management Plans: Opportunities and Problems

Opportunities	Problems
Process provides opportunities for community dialogue and consensus building	Often done in an ad hoc manner or by outside Forestry Staff
Helps a community know the extent of its resources and collectively review	Allows local government a greater hold on how to introduce levels of taxation
Provides written guidelines for forest management	Costly to prepare – May not fit with Working Plans for larger area
Delineates Community Forest Boundaries	May not reflect community management goals
Stipulates Rights and Responsibilities	Planning procedures too complex for communities
Provides a baseline of Forest Stock	Local forest staff may not have capacity to implement
Gives a community a document by which to show their rights to other claimants or stakeholders	Can expose their weakness to implement or protect
Provides a basis for value added activities that might improve trade options	Constraints may be too great to see any point for development
Gains recognition by incorporation into local government land use plans	Can become part of a political agenda rejected by a new incoming local administration

current government laws and legislation, regardless of locally specific forest management objectives and the scale of the commercial operations... Usually these requirements in management plans far exceed minimum necessary information, which the forest managers really need for livelihood-oriented forest management.[197]

An important distinction between forest management planning in commercial operations and community forestry contexts is one of legal tenure and institutional arrangements. In the former, conventional relationships between forest agencies and companies are typically in effect, with little change in the status quo. CF management plans that focus on technical issues, as commercial forest management plans do, may miss the critical questions that require an open and transparent process of debate among stakeholders. By contrast, where community forest management systems are being put in place, an entirely new set of tenure rules and institutional structures may be in the process of being established. As such, the critical issues being addressed in the planning process involve who participates and who is excluded, the territory to be covered, the rights and responsibilities of the CF group (Table 14). These issues generally take precedence over technical questions regarding how the forest will be managed and the process is much more integrative but takes time especially as the roles of the different stakeholders radically shift.

Experience from Africa suggests that the forest management process requires two steps, the first a set of participatory diagnosis and planning activities. The second step is a realignment of power and institutional revision.[198] Clearly, the devolution of forest management rights and responsibilities requires a transfer of authority, an essential step to empower the management plan.

Historical reconstruction of the emergence and development of community forest management systems in eastern India show that local concern over deforestation to effective management may require several decades. Active silvicultural manipulation was not initiated until ten years or more after CFM groups were established and forest protection initiated.[199] In Mae Khan sub-watershed in northern Thailand, hill tribes and local forest department officials have been in dialogue regarding forest management concerns for over 15 years, but only in the past four have collaborative management plans been formulated, and these primarily focusing on the demarcation of boundaries for different use classifications (blending conventional and local cultural forest categories). This planning process has been slowed by delays in clarifying the rights and responsibilities of inhabitants of the upland watersheds as well as the limited capacity of the Royal Forest Department (RFD) to engage the culturally diverse communities in the area. This indicates that management planning is not a one-step process, but an ongoing activity that must focus on the specific issues confronting the community and the forestry agency at that stage in the longer transition process.

The process of state forest agency-community negotiations concerning forest management agreements is sometimes reduced to a simple statement of the terms prescribed by state policies, where communities have little latitude to reject. As Edmunds and Wollenberg note, "Forests have attracted multiple interests, and local forest users are often forced to negotiate with them over forest management practices."[200] This has certainly been the case of some forest-dependent communities in Cambodia and Indonesia, as forest concessions move into their areas. In other cases, forestry field officers, local government officials, development project staff, or NGO workers may periodically engage communities in forest management discussions, conduct PRAs, and hold related capacity building and negotiating activities. Tanaka suggests that unless there are subsequent actions that continue to build towards the recognition and development of community management, these occasional dialogues may lead nowhere. "It is necessary to bring about local dynamics beyond occasional rituals of participation and dialogues with forest users."[201]

An important aspect of the local dynamics is supporting forest-dependent groups, who are often politically weaker than other stakeholders, by allowing them to play a meaningful role in emerging management systems and organizations. Initial management agreements, if they are successfully negotiated, are still fragile and often require renegotiation.[202] Initial agreements are highly vulnerable to challenges from a wide range of stakeholders including community members who disagree, neighboring communities, private enterprises, or the government itself. Agreements may also require change, as new information becomes available regarding territorial issues, rules, rights, or other elements identified in the terms. These observations suggest that forest management plans need to evolve gradually, with flexibility to adapt to learning generated by the community. Tanaka suggests a four-step process of collective forest management planning involving an initial probationary period before permanent handover of forest management rights are transferred, as is the case in Gambia and Tanzania.[203] While some CF laws require a series of steps in transferring forest stewardship authority to communities, most do not. It is a challenge for policy makers to find ways of supporting a sequential process of devolution that supports capacity development and a gradual and growing formalization of rights and increasingly defined management plan.

Royalties, Taxes, and Contractual Agreements

As Edmunds and Wollenberg note, taxes and regulations are an integral part of forest governance and can function in both private sector and community management systems to ensure that revenues are generated to ensure public interests and ecological services are protected.[204] It has long been observed that forest department impositions of taxes and regulations on community-level harvesting, transport, and trade in forest products can undermine local forest initiatives. Obtaining permits for harvesting and transportation can be a laborious and time-consuming process that may dissuade communities from planting trees. It may also be an opportunity for corruption. In some cases, illegitimate harvesting is permitted through payments between elites and forest department officials, while legitimate harvesting is prevented due to such barriers.[205] Experience in recent decades from India, Indonesia, and other nations indicates that little evidence is available to show that forest royalties, taxes and fees have resulted in significant investments in forest restoration and protection, while there is a growing body of empirical information that indicates such regulations have distorted forest product markets, created disincentives among rural families to manage forests, and contributed to corruption.

Contractual agreements, that frequently represent the legal side of collaborative forest management programs, have also been widely criticized for their tendency to protect the authority of the forest department, while extending more responsibilities than benefits to communities.[206] While such contracts are intended to provide communities with greater tenure security, and often do, they have also been critiqued in

TABLE 15: Ratio of State Forest Area to Forest-Dependent People, Forest Department Staff and Budget

Data as of 2002	Cambodia	Indonesia	Philippines	Thailand	Vietnam
Total Official Forest Land ('000 ha.)	10,535	135,868	15,882	30,143	10,915
Actual Forest Cover as of 2000	9,335	104,986	5,789	14,762	10,915
Inadequate cover* ('000 ha.)	1,200	30,882	10,093	15,381	na
Forest-Dependent People ('000)	1,800	65,000	24,000	11,000	27,000
Forest Land Per Forest-Dependent (hectare/person)	6.0	2.10	0.70	2.80	0.40
No. of Forest Department Staff+	858	17,697	23,373	16,072	10,200
Forest Land Per FD Staff (hectare/staff)	12,300	7,700	680 (4,000)	1,870	1,070
Budget (in thousand of US$ and €)+	$608 €742	$147,846 €180,300	$97,334 €118,700	$181,532 €221,380	$57,133 €69,674
Budget per hectare of forest land (in US$ and €)	$0.06 €0.07	$1.09 €1.33	$6.13 €7.47	$6.02 €7.34	$5.23 €6.38

*Inadequate Cover = Official Forest land – Actual Forest Cover
+Communications with AFN country partners: Ly Chou Beang (DFW, Cambodia), Suhardijono (DepHut, Indonesia), Dionisio Tolentino (DENR, Philippines), RFD website (Thailand), Ho Manh Tuong (FIPI, Vietnam); €1 =US$1.2

terms of their tendency to reflect a forester's vision regarding appropriate management, their promotion of state interests over communities, and, in some cases, the disproportionate authority of the state to void them without recourse.

Lack of Government Resources

While decentralization has brought new authority and resources to many local government institutions in Southeast Asia, this process is still in an early phase of implementation and capacity is limited in a number of ways. Local government officials typically focus on critical infrastructural needs including roads, water, health care, and education, leaving natural resource management concerns as secondary issues. Local elected representatives frequently have little experience with NRM issues, typically less than their rural constituencies. While some authority is being given to local government to oversee resource management, they have minimal budgets or staff to implement activities. The role of local government will very likely grow in the future, as water resource problems grow in scale.

Forest departments are increasingly acknowledging that their present human and capital resources cannot bear the entire responsibility for managing millions of hectares of forest lands. Forests are decreasing while the rural population is increasing, and at the same time forest department budgets have remained relatively stagnant. In Southeast Asia, the estimated ratio of forest land to forest department staff ranges from 2,000 to 12,000 hectares per staff (see Table 15).

Only a modest percentage of total forest department personnel operate at the field level, so the ratio is actually much higher. On the other hand, the ratio of forest land to rural population ranges from one to two hectares per capita reflecting the predominance of community presence.

In the Philippines, the Haribon Foundation estimates that on the ground there is only one guard for every 4,000 hectares of forest. In contrast, the Philippines has approximately 24 million people living in rural areas who, given adequate incentives, can potentially take responsibility for forest management. Indonesia is estimated to have one forester in the field for every 100,000 to 300,000 hectares of forest land.[207] Ministry of Agriculture and Rural Development (MARD) staff in Vietnam are all based in Hanoi and operate through the Forest Protection Branches under the District People's Committees.

Financing forest management on the ground is also a severe problem. While central government budgets for the forestry sector may average from $ 1 to 8 per hectare, most of these funds are utilized on core support costs for forest departments including offices, staff and transportation. The amount of funding for forest protection and management is a fraction of this. In the Philippines, the Haribon Foundation estimated that funds actually earmarked for forest protection was only $0.02 (P1.50) per hectare. Forest guards in the Philippines earn approximately $100 per month, making them very vulnerable to accept bribes for illegal logging activities. In Cambodia and Indonesia, the field staff salaries are much lower and budgets for protection are considerably less than the Philippines.

The human and financial constraints faced by forestry sector planners in the Southeast Asian countries is creating a growing awareness that community involvement in forest protection and management is not simply an option, but a necessity. Yet, there remain many traditional foresters who seek to retain authority and control over forest revenues. This dynamic has both encouraged the spread of CF, but also restricted its implementation and its meaningful empowerment. Unsupportive attitudes and limited knowledge of community forestry among forestry officials also present a human resource constraint to program implementation. In Cambodia, Indonesia, Laos, and Vietnam, community forest management is a recently introduced concept for foresters and the transition in staff attitudes is at an early phase. Some older staff see community forestry as a distinctly different approach than what they were trained to do, an approach that places greater authority in the hands of rural people, and on that may erode their own authority.

In Thailand and the Philippines, the Royal Forest Department and the Department of Environment and Natural Resources have expended considerable resources for staff training programs regarding community forest management over the past decade. In those countries, agency staff is generally well-trained, with most officers at the provincial and district level holding college degrees. Nonetheless, despite efforts to support transitions in staff orientation from conventional modes of commercial timber towards participatory approaches, staff attitudes have adjusted slowly. An RFD community forestry training specialist estimates that the percentage of mid-level watershed managers who accept CF as a viable approach has increased from 10 percent in 1990 to, perhaps 50 percent today. Yet, he acknowledges that only a small minority may be active advocates of community forest stewardship.[208]

Disempowered Rural Communities

Incorporation in mainstream systems is best indicated by availability of basic services in an area. Even in rural areas with significant populations, basic services such as water systems, health care, schools and roads are not always available. The lack of services has marginalized many of the communities and constrains the implementation of activities developed under CF programs. In several countries, such communities are not able to access government resources because old policies continue to view them still as squatters on state lands. A population that is not recognized in terms of tenure and social services then has an empowerment problem. Most hard hit are the minority communities that may not have political acceptance or even recognized citizenship. For the most part, remote forest-dependent communities face an uphill struggle in obtaining government programs as they are neither a threat to more totalitarian regimes, nor an important part of the electorate in democratic systems.

CHAPTER 15

Contexts for CFM Application

Chapter 15 explores potential applications for community forestry in Southeast Asia in the future. Community management strategies are already being used to restore closed timber concessions, rehabilitate degraded watersheds, and regenerate coastal mangroves and flood forests. Communities are gaining recognition for the important role they play in managing millions of hectares of agro-forests throughout Southeast Asia. More recently, in a growing number of protected areas and national parks in Southeast Asia communities are beginning to be formally integrated through different forms of co-management, yet community contributions to management of many forest environments remains as potential still to be unleashed.

Restoring Logged-Out Timber Concessions

Many Southeast Asian nations have already phased out export-oriented industrial logging including the Philippines, Thailand, and Vietnam, or are scaling down logging operations such as Indonesia, Cambodia, and Malaysia. In each country, millions of hectares of logged forest land, once under private sector timber concessions, must be placed under new management either directly through a government or other entities, as their leases are cancelled or come to an end (see Table 16). Former timber concession areas can support new orientations in community management. In the Philippines, after the enactment of the new constitution in 1987 driven by the People Power Revolution, the number of timber license agreements was reduced from 159 in 1986 to less than 20 in 1998.[209] By extrapolation, this policy change has opened up 8.7 million hectares, or 58 percent of the total forest lands of the country, to other forms of forest management such as CFM.

In 1978, Indonesia had 41 million hectares of forest land under timber concessions. By 1996, this increased to 70 million assigned to 692 concessionaires. Since that time, due to international pressure and increasing local unrest, the concession area was reduced to 49 million in 1998, and the number of concessionaires decreased by half. The Indonesia government plans to further reduce the number of concessionaires to 50 units before 2010.[210] If this target is followed by a corresponding decrease in areas under forest concessions, then CFM can play a major role in these areas.

Cambodia adopted the timber concession system in the early 1990s and awarded 6.5 million hectares to 30 logging enterprises between 1994 and 1997. International and donor pressure to reform the forest sector prompted the government to cancel 4 million hectares under concession which were found to be uneconomic, representing two-thirds of timber area, and declare a moratorium on any further commercial felling

TABLE 16: Areas Under Timber Concessions

Timber Concession (in 000's hectares)	Cambodia*	Indonesia***	Philippines***	Thailand+	Vietnam++
Timber concession area (past)	6,500 (1994-97)	70,612 (1996)	10,000 (1970s-80s)	217 (1977-81)	4,000 (1980s)
Timber Concession area (present)	2,500 (2004)	49,257 (1998)	1,300 (2000)	151 (1992-96)	3,578 (1999)
Additional Areas Potentially Available for CFM	4,000**	21,355	8,700	66	422

*Past timber concession area (1994-97) from Prom Tola and McKenney, Bruce. Natural Resources and Rural Livelihoods in Cambodia: A Baseline Assessment. CDRI, 2002, p. 75. and McKenney, Bruce. Baseline Assessment of Cambodia's Forest Sector. WWF, 2001, p. 15. ; Present timber concession area (2004).
** Figure for 1996 from Asia-Pacific Forestry Sector Outlook Study: Country Report - Indonesia. FAO Rome, 1998. Area covers 483 concessionaires operating in natural production forest lands. Per email communications with Pak Suhardijono of Ministry of Forests, timber concession area as of 1978 is 41.47 million hectares; Figure for 2000 from Forest Watch Indonesia / Global Forest Watch. State of the Forest Indonesia, 2002. p. 28.
*** Forest Out of Bounds, FAO 2001, p. 116
+ Forest Out of Bounds, FAO 2001. Represents areas assigned to RFD and private sector for reforestation (p. 177); area under timber concession could be higher. There were 22 companies granted logging rights in designated national parks and wildlife sanctuaries in the 1980s, before the Bangkok flooding in 1988 (p. 170). Forest land during the 1970s was leased to private sector for 62.50 baht/hectare/annum on 30-year plantation contracts.
++FIPI Table of Forest Management Allocation. Area managed by State Forest Enterprises.

until acceptable management plans had been approved by the Forest Administration. Many of the concession areas were largely logged out by the time their leases were cancelled, and have subsequently been reclassified as shrub lands. [211]

While tremendous potential exists to engage communities in managing millions of hectares of logged-over timber concessions, little progress has been made on the ground in developing formal agreements with resident villagers. There are exceptions where donor agencies have funded experiments with community-based timber operations including Sanggau, Indonesia (GTZ), Fomacop – Laos (World Bank), and in southern Philippines (USAID); however, their strategies placed greater emphasis on small-scale logging, rather than forest restoration.

Rehabilitating forests that were commercially logged during the 20th century presents a major opportunity for CF in the 21st century. Vast areas of logged-over forests have been badly disturbed, experiencing watershed deterioration and soil erosion from logging roads and inefficient extractive activities. For the most part, industrial timber companies and national governments have done little to mitigate the impact of logging activities in terms of dealing with fuel load build-up from slash, thinning of dense re-growth to facilitate tree growth, or restoring damaged riparian arteries. Forest resident communities are well-positioned to play an important role in restoring the ecological health of these environments.

Rehabilitating Degraded Watersheds

Increased occurrence of flooding, landslides, and water shortages in the lowlands illustrates the importance of watershed management. In many Southeast Asian countries, between one-quarter and one-third of all forest land is formally designated for watershed protection (Table 17). Evidence has shown that declaring watersheds as national reserve forests or protected areas is not enough to stop their degradation. In Thailand, RFD classifies the 9.32 million hectares of watershed reserves under Watershed Class 1, further divided into 2 sub-classes. Watershed protection forests (sub-class 1A) include the headwaters of rivers, usually located at higher elevation, and steep slopes, and should remain as permanent forest cover. Disturbed watershed protection forests (sub-class 1B) are areas with similar physical and environmental features as Class 1A, but parts of the areas have been cleared for agriculture or have been settled. Watersheds falling under this sub-class require special soil conservation measures and thus need to be replanted as forest or maintained under permanent agroforestry systems. Government can collaborate with communities in regenerating vegetative cover for soil conservation in these areas.

In Indonesia, 11 million hectares of degraded state forest land are under protection for soil and water stabilization and requires reforestation. Another 20 million hectares of old logging concessions continue to be classified as Production Forest, but are located in watershed areas and need rehabilitation. The Philippine government had 1.5 million hectares under watershed forest reserves as of 2001. Most of these areas are not under effective on-site management systems. Some local governments are taking a closer look at how these reserves are being managed. The provincial government of Nueva Vizcaya, for instance, collaborated with the regional office of DENR in establishing on-site forest land management for the Lower Magat Forest Reserve, the water source for the dam supplying central Luzon rice plains. Nueva Vizcaya's approach was to secure the tenure of settlers within the watershed through land management agreements as an incentive for them to act as a "social fence" against further encroachment and help repair the

TABLE 17: Areas Under Watershed Reserves

In '000 hectares	Cambodia*	Indonesia**	Philippines***	Thailand+	Vietnam++
Watershed Reserves	1,346	50,270	1,500	9,320	5,350
Total Forestland	10,535	135,868	15,882	30,143	10,915
Watershed Reserve as a % of total forestland	13%	37%	9%	31%	49%

* Protected forests under the jurisdiction of MAFF based on DFW statistics as per email communications with Mr. Ly Chou Beang
** Forests with watershed protection as first priority based on MoF statistics as per email communications with Ir. Suhardijono
*** DENR-FMB, 2001. www1.denr.gov.ph.
+ Data for Watershed Class 1 from RFD 1998 and Green World Foundation 1999 (cited from Forest Out of Bounds, FAO 2001, p. 168.)
++ Data from MARD 2000 (cited from Forest Out of Bounds, FAO 2001, p. 186-187.) Refers to protected forests (forests for coastal and watershed protection).

degraded watershed. In Maasin, the watershed of Iloilo City on Panay Island, communities that have been granted a CBFM agreement are getting support from the city government for their protection activities.[212]

While effective watershed management is essential for ensuring important environmental services, the simple demarcation of upland areas in national capitals will not achieve this goal. It must be linked with on-site management that can guarantee that these regions are strictly conserved and utilized in low-impact ways. Local communities appear to be one of the most cost-effective ways to ensure on-site management, both because they can act as a barrier to watershed disturbance by outside actors, whether they be private sector firms, land speculators, or migrant populations, and as they can cause problems directly through clearing forest cover to open new areas for settlements and agriculture. Without the active involvement of local communities in watershed planning and management, conflict with local villages is inevitable, further undermining management efforts. While some progress is apparent in Southeast Asia in linking watershed residents into management systems, these efforts are limited and remain largely at a pilot project phase of development. Where these watersheds are already degraded planting with fruit trees rather than pulp wood species is more sustainable. In the decades ahead, community-based watershed management needs to be broadly implemented in much of the region.

Regenerating Coastal Mangroves and Flood Forests

The forest sector has historically focused on management issues in environments with high commercial timber values, such as lowland tropical evergreen forests. Community-based forest management is typically discussed in the context of the uplands and degraded lowland deciduous forests. Mangrove, swamp, and flood forests, which do not have much commercial wood production value, apart from fuel wood and charcoal for domestic consumption, have been given relatively little attention. As the concept of forest management and forestry broaden, and as scientists increasingly identify the important environmental services provided by aquatic forests, interest in these ecosystems has increased. Table 18 indicates the current area under aquatic forests, representing a fraction of the area mangrove and flood forests covered fifty years ago. Since the mid-1980s alone, it is apparent that Indonesia, the Philippines and Thailand have lost more than half of their mangrove forests.

In the past decade, however, the role of communities in protecting and restoring edaphic forests is starting to gain greater attention, with some countries like Cambodia and the Philippines, already actively engaged in such a management transition.

Much of Cambodia's population relies on fish for protein caught in the Tonle Sap. But the rapid decline in flood forests over the past 30 years has contributed to the decline of fish stock in the Great Lake. This has generated alarm within the government and the international community that views the flood lake as a rare ecosystem the world cannot afford to lose. In 2002, the Prime Minister transferred 56 percent of the total area under commercial fishing lease management, totaling over 500,000 hectares to local communities living around the lake. A Community Fisheries Law is currently under consideration by the legislature. Community forestry projects covering over 40,000 hectares of mangrove forests along Cambodia's southern

TABLE 18: Mangrove and Flood Forests

In '000 hectares	Cambodia *	Indonesia	Philippines	Thailand	Vietnam
Mangrove and Flood Forests 1980s+	250	4.413	1,000	2,000	NA
Mangrove and Flood Forests 2003 @	515	2,930	109	244	104
Total Forest land	10,535	135,868	15,882	30,143	10,915
% of total forest land	5%	2%	1%	1%	1%

+ Collins, Mark N. Jeffrey A. Sayer, and timothy C. Whitmore The Conservation Atlaas of Tropical Forests: Asia and the Pacific (New York: Simon & Schuster, 1991).
@ FAO State of the World's Forests 2003 except for Cambodia.
* Hong-Narith (DFW). Asia-Pacific Forestry Sector Outlook Study: Country Paper on Some Aspects of Forestry in Cambodia. FAO Rome, 1997. Includes 481,224 hectares of flood forests, 17,655 hectares rear mangrove forests, and 16497 hectares mangrove forests.

coast are also under way, reversing mangrove deforestation trends.

In Indonesia, mangrove forests play an important role in harnessing fish stocks and protecting the country's vast coastlines. Mangrove forests have experienced widespread degradation in recent decades, particularly along the east coast of Kalimantan, Java's north coast, the west coast of North Sumatra, Aceh, Bali, and South Sulawesi. While the National Movement for Land Rehabilitation (*Gerakan Nasional Rehabilitasi Lahan* / GNHRL) and other local community initiatives in Sulawesi have conducted efforts to rehabilitate mangrove forests, these have not been not done in conjunction with any formal transfer of official land management rights to those involved in the movement.

The community-based forest management framework of the Philippines has extended its application to mangrove forests. Mangrove forests have been severely denuded over the past century with mangrove areas declining from 500,000 hectares in 1920 to less than 100,000 by the mid-1990s. People's organizations living around coastal areas, usually composed of fishing families, are increasingly seeking tenure rights through the submission of coastal resource management plans to DENR. Approval of community management plans will secure their continued access to the area for 25 years. Compared to upland communities, it was easier for coastal communities to see the benefits of their management activities. Mangrove forests show substantial regeneration after five years, considerably faster than most upland forests, and their re-establishment brings immediate rewards in the form of increased fish stocks, crabs, and fuel wood. The rapid flow of benefits from improved management reduces the need for external subsidies from government and donor agencies to catalyze mangrove protection initiatives. Community involvement in aquatic forest restoration and management is also taking place in the Mekong Delta in Vietnam, where degraded mangrove forests have been regenerated over the past decade.

After the catastrophic tsunami of December 2004, attention has been drawn to the role of coastal forests, especially mangroves, as buffers against wave action as well as major storms. Many of the areas most affected by the tsunami had historically possessed coastal mangrove forests that had been cleared in past decades for fish ponds and fuel wood. These areas include the eastern coastline of India, northern and eastern Sri Lanka, as well as along the coast of northern Sumatra and the western Malayan Peninsula. CFM has tremendous potential in flood and coastal forests, provided that supportive policies that extend tenure authority and management rights to local villages are in place. Restoration of mangrove forests will not only help protect coastal communities from natural disasters, but rehabilitate fisheries central to local economies.

Protection of National Parks

Media promotion of nature conservation, eco-tourism, and biodiversity has generated considerable interest among urban populations throughout Southeast Asia, creating broad-based support for protected areas and national parks. The amount of protected areas in Southeast Asia varies widely among the nations of the region, ranging from less than 10 percent of forest lands in Vietnam, to over 75 percent in Thailand (see Table 19). Aside from protecting critical watersheds, national governments view protected area systems as an opportunity to generate revenue and foreign exchange from local and foreign tourists. International conservation organizations are also effectively lobbying to set aside additional areas for conservation, and raising funds to facilitate the establishment of new protected areas. While the creation of a national environmental awareness is a valuable complement to civil society in Southeast Asia, frequently the needs of forest-dependent people are ignored or not given the attention they deserve. Part of the problem is that there has been a tendency in Southeast Asia over the past three decades to adopt a "North American" conservation model. This approach assumes little or no resident population within nature conservation areas, unlike the "European" conservation model, where local communities are often an integral part of the park. The North American model may require resettlement, or at the least, a minimal role for communities in management activities.

In Thailand, which has tended towards the exclusion of local populations living near or inside protected areas, 76 percent of the nation's forests are classified as conservation areas, with approximately 5 million hectares of national parks. The conservation effort is consistent with Thailand's goal of bringing 40 percent of the land area under forest cover, and also reflects the strength of urban middle-class nature conservation advocacy groups. The steady expansion of conservation areas over the past decade has brought insecurity to hill tribes and other forest-dependent people in some areas, and has increased the rift between upland communities and protected area managers. Thailand's northern watersheds are currently under the management of the Protected Area Department, within the recently formed Ministry of Natural Resources and Environment. With 11 million forest-dependent people, the government agency seeks to find a politically viable way of implementing national conservation policies, without marginalizing and displacing a sizable proportion of the rural population.

Upland watershed communities in northern Thailand are increasingly organizing, with over 1,000 villages currently participating in networks that seek to engage government in a policy dialogue regarding their tenure and forest rights. Co-management systems between upland communities and park managers need to be developed to avoid social conflict and to achieve environmental management goals. Currently, the Sustainable Development Foundation is building capacity of communities and park officials in participatory zoning in Ob Luang National Park and five other protected areas. While successful pilot projects are being implemented in a number of northern watersheds in Thailand, the implementation of co-management systems has been slowed by the failure of policy makers to approve a viable community forestry bill and by conservation policies and protected area management plans that fail to provide resident populations with opportunities to participate in management decision making.

In the Philippines, people inhabit virtually every protected area. Resettlement to achieve strict protection is not a viable option due to the scarcity of land for relocation, lack of financial support, and political resistance. Therefore, the NIPAS Act was promulgated and implemented to facilitate co-management and multi-stakeholder participation. Each protected area is required to establish a workable Protected Area Management Board (PAMB) to implement the General Management Plan. According to the Act, DENR, local government, and communities must develop a management plan, and all stakeholders should have representation in the PAMB. The protected area is zoned into strict protection and multiple-use. The zoning process becomes the mechanism through which an agreement is reached regarding areas that can be used by communities, and which areas they will help protect. Though the framework for dialogue and collaborative management is sound, numerous problems have been experienced in its implementation.

Similarly, in Indonesia communities reside in almost all protected areas (national parks, protection forests, nature reserves, and recreational parks). In the past, there were efforts to resettle these communities outside the conservation zones, but by the 1990s such initiatives were halted due to insufficient funds and community reluctance to move. Co-management programs are now being explored to accommodate the socio-cultural situations of communities.

In most protected areas in Southeast Asia, community involvement in conservation areas is not an option, but an imperative if the goals of biodiversity conservation are to be achieved. Unfortunately, progress towards meaningful engagement of local people in field activities related to conservation and in decision making regarding use, has been slow and limited to a relatively few pilot areas. Aside from constraints on human and financial resources, the implementation of collaboratively managed protected areas has been restricted by the continued adherence to top-down park

TABLE 19: Protected Areas

in '000 hectares	Cambodia*	Indonesia**	Philippines	Thailand+	Vietnam++
Protected areas	3,402	49,500	2,509	23,037	1,550
Total Forest land	10,535	135,868	15,882	30,143	10,915
% of total forest land	32%	36%	16%	76%	14%

* Hong-Narith (DFW). Asia-Pacific Forestry Sector Outlook Study: Country Paper on Some Aspects of Forestry in Cambodia. FAO Rome, 1997. Includes areas classified as natural parks, wildlife sanctuaries, protected landscapes and multiple-use areas. Flood forests around Tonle Sap are considered as protected area. www.fao.org.

** Ministry of Forestry. Asia-Pacific Forestry Sector Outlook Study: Country Report - Indonesia. FAO Rome, 1998.

*** Forest Out of Bounds, FAO 2001, p. 107.

+ RFD website, 2003. Data as of 2000.

++ Forest Out of Bounds, FAO 2001, p. 186-187. Refers to special use forests (national parks, nature reserves, and cultural sites).

management models that allow little room for community engagement.

As a consequence, many of the newly created protected areas have limited on-site protection and are frequently subject to illegal logging, poaching, and encroachment. The future quality of the region's protected areas will, to a large extent, depend on the success of policies and field programs that mobilize local communities to meaningfully collaborate in managing protected forests. Linking modern conservation values and systems with traditional cultural forms may be an effective mode for creating new management partnerships for protected areas.

Managing Forest Gardens and Agroforests

Many agricultural landscapes in Southeast Asia can be classified as some form of agroforest or forest garden, a land use that combines "planted trees" with forest flora and fauna, either retained or naturally regenerated vegetation.[213] Frequently, agroforests are a phase of a broader agriculture cycle that includes mixed horticulture. Tree planting in these agroforests can occur in an open field stage, often in between food crops, or in small gaps or clearings in existing forest.[214] Agroforestry practices that operate throughout the region range from traditional forms to newly emerging sustainable forms of more intensive production. There are no precise figures yet on the extent of "agroforest" areas in Southeast Asia, as different data sets are maintained by agriculture and forestry agencies. Nonetheless, agroforestry systems are extensive throughout the region.

In Indonesia, agroforestry systems are complex and include the Javanese practice called *wono dusun* on private land, and the *kebon-talun* in West Java. For centuries, the Javanese and Sundanese have designed these multi-storied agroforests and forest gardens to meet an extremely wide range of subsistence needs, with each plot possessing dozens of species of trees. Many such "people's forests" have a three-layer canopy structure involving a variety of tree species intercropped with a mixture of climbers, shrubs, annual food crops, and herbs. Over the past two hundred years, external markets have influenced the species that Javanese introduce into their agroforests. However, the structure has always retained substantial canopy closure and ground cover allowing for excellent soil and water conservation, as well as species diversity. Agroforests have demonstrated a capacity to sustain the productivity of the land, as well as support a healthy local economy. By utilizing their local knowledge and retaining a high diversity of species, households constantly adapt the composition of their agroforests to respond to markets, pests, and disease allowing the necessary flexibility for greater sustainability.

Good markets for tree products such as fruits, resins, and latex have allowed a transition of substantial areas of Southeast Asian forest to agroforest. As the need for cash income increases along with an increase in land scarcity, a pattern widespread in Southeast Asia, many traditional swidden farming communities are shifting their swidden areas from long fallow cycles into permanent agroforestry production. While some government programs and donor projects are encouraging such transitions in forest management systems, the widespread transition towards agroforestry and forest gardens from swidden is primarily market driven. Private sector investors and land speculators often provide capital for such land use transitions, though indigenous people may see their ancestral lands alienated in the process. Such trends are apparent in northeast Cambodia, the central highlands of Vietnam, Northeast India, and many other regions.

Home gardens, fruit orchards, cut flowers, sloping agriculture land technology (SALT), and trees around fields are some of the contemporary agroforestry systems that evolved from these market opportunities. The migrants in Mindanao are integrating traditional elements of their multiple storied agroforests into their SALT-based farming systems. In Vietnam, a system of forest gardens that integrated aquaculture and animal husbandry, *voun ao choung* (voun =garden/orchard; ao =fishpond; choung = piggery/poultry), was traditionally practiced in the Red River Delta. This system is being promoted by VACVINA to families, schools, health centers, factories, offices, day care centers and cooperatives in areas where government granted land use certificates. In northern Thailand, farmers nurture *miang* tea agroforests by planting the tea species in the understory of natural forests, gradually modifying the floral composition without a clear felling stage. [215]

While agroforestry systems are widespread in Southeast Asia, forest cover statistics based on satellite imagery tend to include them as part of the national forest area. The Indonesian Ministry of Forestry data on actual forest cover based on satellite imagery reveals that of the 93.5 million hectares of State Forest Zone analyzed in 2002 (excluding West Papua), 52.5 million had forest cover, 27.6 million had no forest cover and that for 13.3 million the quality of the images were insufficient for accurate interpretation.[216] While there was no attempt to differentiate between agroforests, or groups of trees planted by local people on land they claim as their own, and natural forest, it is likely that large areas of land identified in the above analysis as "forested," are in fact agroforests with communal or private rights attached.[217] It is possible to estimate that at least ten to fifteen million hectares of land are under productive multi-storied forest gardens across Southeast Asia. A recent report by FORRESASIA, a research program that has been exploring forest culture for over a decade, notes that cultivated forests cover from 6 to 8 million hectares in Indonesia. In assessing the economic importance of cultivated forest to the rural economy, the report notes that:

> *In Indonesia, they provide 80 percent of the processed and exported rubber latex, 80 percent of the dipterocarp resin and 95 percent of the benzoin resin traded in and outside the country, between 60 to 75 percent of the main tree species (clove, cinnamon, nutmeg) produced for national and international markets, roughly 95 percent of the various fruits and nuts marketed in the country, as well as a significant part of the bamboos, small cane rattan, fuel wood, handicraft material and medicinal plants traded or used in the country.*[218]

The amount of land potentially available for conversion to forest gardens and agroforestry systems could be at least 50 to 60 million hectares throughout Southeast Asia. Further, it is apparent that land under cultivated forests and agroforestry is increasing steadily as swidden farmers convert land once under long rotation of subsistence food crops to permanent, more market-oriented tree crops. The value of commercial fruit, resins, spices, fibers, and medicinals that would be produced through such a landscape transformation would likely exceed $100 billion annually at current market prices, not including valuable environmental services that could be secured.

The multi-structure canopy of many Southeast Asian agroforestry systems mimics the natural forest ecosystem and contributes to conserving biodiversity and other environmental services as natural forests. A recent study found that in Krui, South Sumatra of Indonesia, local agroforests (*damar* forest gardens) possessed more than 50 percent of the tree and bird species as old-growth rain forests within a neighboring national park, as well as providing habitat for endangered mammals including the Sumatran rhinoceros, tigers, tapir, and gibbons. By contrast, local rubber estates possessed less than 5 percent of the same species.[219] Many agroforests, due to their diverse species composition provide rural households with a secure food source during times of shortage. The variety of cash crops found in many agroforests buffers rural families from common price fluctuations among such commodities as cloves, coffee, pepper, and others traded on international markets. Government's provision of tenurial rights to people's forests (*hutan rakyat)* has proven to be a cost-effective approach in maintaining vegetative cover on upland areas in Java and the Outer Islands of Indonesia.

Agroforestry systems, due to their diversity and dispersion receive relatively little policy and program support from government, especially when compared to oil palm, coffee, and rubber estates. In Southeast Asia, agroforests are primarily production systems for low-income, small landholders, while past government support in the forestry sector has focused on larger commercial timber and plantation initiatives. While agroforests are often far more environmentally sustainable and economically productive than other modes of production within the forestry sector, it can be argued that they have not received significant government support in the past because it is difficult for powerful interest groups to extract revenues from them. Supporting community management by encouraging the extension of agroforestry systems and forest gardens on public land is one way to respond to the increasing need for greater acknowledgement of peoples' rights to the resources and way of life. At the same time, it is one of the most promising ways to reforest and afforest degraded lands and return them to economic productivity, while restoring their environmental service functions.

PART VI

Prospects for Change

The impact of community forest management can be enhanced through the effective implementation of CF policies and programs. Well-articulated national laws and strategy statements can facilitate the transition from isolated pilot activities and indigenous practices to a broader incorporation of rural people in formally recognized systems of forest tenure and stewardship. The political commitment of government to move forward with resource management transitions, often in the face of opposition from vested interests within government and the private sector, has been a key element in the national expansion of CF in the Philippines, and its absence in Indonesia has limited the impact of CF in that nation. That ability of CF strategies to improve environmental and socio-economic conditions is also determined by the capacity of line agencies and NGOs to implement CFM policy agendas and new legislation and the willingness of development agencies to provide financial and technical support. Finally, the effectiveness of CFM laws, policies, and program in terms of their capability to conserve forest cover and improve rural livelihoods will be determined by the knowledge of, interest in, and capability of community leaders, members, and institutions to relate to and implement them.

National strategies that are funded and flexibly designed to respond to local social and economic conditions appear to accelerate and help sustain the implementation of CF-related legislation and policy. By contrast, national programs that are designed more in response to the concerns of forestry agencies, private sector interests, and other stakeholders beyond the community, may experience limited local support and may be difficult to sustain after project subsidies end. For example, some community forest management agreements have been designed by government officials and project staff to enhance timber extraction on scales that exceed community management capacities, while failing to address community needs and forest management goals. Other agreements may emphasize protection and conservation, while giving little attention to income generation, despite poverty within the community. Neither strategy adequately responds to most community conditions and needs, but rather dictates an agenda being devolved from the center. Such problems are common in situations where government goals and development agency priorities are the driving force between national efforts to draw communities formally into forest management. Such agreements generally require reassessment within the first few years of their ratification, especially if communities are allowed and assisted to articulate their own approach to management.

Throughout Southeast Asia, many nations are approving policies that devolve greater authority to local government. Decentralization can result in gradually increasing local governance in development and natural resource use planning decisions, especially at the district and sub-district level. While the inclusion of NRM responsibilities in decentralization legislation is increasingly common in Southeast Asian countries, delegating significant control over forest product revenues and management budgets is rare.

Decentralization and devolution laws and policies may be associated with greater involvement of communities in resource management, though this is not an inevitable outcome. In India, throughout the 1970s and 1980s, local *panchayat* governments were routinely given responsibility for social forestry projects. The experience was generally disappointing, however, as panchayats with five to twenty member villages were often too distant from forest areas to manage them effectively, while leadership was frequently unprepared or disinterested. In many Southeast Asian countries, sub-district level administrators are already overwhelmed by numerous tasks such as the provision of such basic services as health care, schools, roads, and civil administration. Natural resource management represents a new and complex set of duties for which time, skills, and budgets are minimal. Some local government officials, however, recognize that they must harness community abilities and interests to address the needs for forest, land and water management.

Government initiatives driving decentralization and community forestry, while sometimes overlapping, have generally been triggered by different concerns, setting forth distinctively different goals, and as a consequence are rarely designed for smooth and integrated implementation. Decentralization often reflects political power balancing between the center, province, district, and sub-district, often responding to pressure from local political forces. Most CF policies, by contrast, are formulated by the center in response to pressure from development agencies for forestry sector reform and by local community demands. In the Philippines, for example, CF policies were developed independent of the processes of local government planning. Countries like the Philippines are seeking ways to integrate CBFM within its emerging land use planning systems. In Vietnam, especially in the north, the commune administration has considerable authority and capacity to address development issues, including resource management and planning, and often effectively engages village leaders and members, allocating responsibilities accordingly, often based on the customary practices of member villages. Commune planning activities (SEILA) held in Cambodia over the past few years indicate that one-third of the local governments viewed community forestry as one of their development priorities. Donors are currently exploring ways to make small $1,000 block grants to interested communes to support CF.

In a recent study of three country experiences with devolution of NRM, including China, India, and the Philippines, a large team of researchers coordinated by the Center for International Forestry Research (CIFOR), concluded that: "...devolution policies have often done little to help local forest users shape what forests look like, protect and promote their livelihoods, or establish and maintain their own decision-making processes."[220] The CIFOR team surmises that this lack of impact in terms of empowering community resource management reflects divergent interests between government and communities including different visions of forest quality, forest benefit distribution, and rights of management decision making. The study concludes that devolution policies represent a "shift in the manner in which central government controls forest management, rather than a genuine shift in authority to the poorest forest users."[221]

The CIFOR report notes that timber remains a key goal of management, even under devolution programs, with conservation a second goal where environmental values are high. While devolution policies vest communities with new responsibilities for forest regeneration and protection, government retains authority over important aspects of management including planning decisions and forest product flows. Government does this through taxes and regulations, and contractual agreements, and through co-opting local organizations.[222] The CIFOR assessments from China, India, and the Philippines present a rather negative view regarding the impact of policies and programs that are devolving forest management to communities and local government, with ample justification. Yet, it can also be argued that the devolution policies of the past decade are a starting point in a new round of negotiations between the center and the people. For much of the past century, the forests of Asia have been usurped by the state, at least on a legal level. Technocratic agencies have gradually taken complete legal control over the resources, and only recently have begun to "devolve" some of the rights and responsibilities of management downwards. That they have taken this step at all is a measure of progress, and one that has been welcomed by communities in many areas.

As some of the authors of the CIFOR study note, "Despite their reservations about JFM, many villagers considered it strategically important to gain some formal recognition from the Forest Department."[223] From the perspective of the past fifty to one hundred years, many forest communities have gone from being viewed as "forest encroachers," "illegal forest occupants," "destroyers of the forest," and "criminal tribes," to being perceived as management partners and forest protectors. While this shift may not yet be reflected in significant management rights, it is a major change in attitude on the part of government and provides a basis to lobby and organize politically for greater future authority. Attitudinal change and bureaucratic reorientation is time consuming, and the transfer of authority is a process that most government agencies do only reluctantly and only when under pressure. This pressure may initially come from above in the form of development agency conditionalities, but gradually it is building from below in the shape of an increasingly influential rural political movement within gradually democratizing societies. It is this process that could eventually enable a meaningful shift in resource control.

As early as the 1970s, some Asian countries were initiating government-sponsored social forestry projects, engaging households in farm forestry and village woodlot activities. By the late 1980s, community forest management had grown as a development strategy, manifesting itself in a diversity of projects and programs. However, there were few countries with policy instruments to legally support the role of communities as managers of state forest land. As Part III documents, the legal and policy framework endorsing CFM has developed gradually during the 1990s, and continues to unfold.

The process through which legal frameworks are formulated to enable community engagement in state land management varies widely from country to country, depending upon the political environment, government and legal institutions, and influence of donor agencies, among other factors. In many cases, it begins with projects and responses to critical events, and evolves into national programs with policy support articulated in the form of regulations, government orders, and amendments to forestry laws. Countries such as India, Philippines, and Nepal have over a decade of experience in CF policy development and implementation. India started implementing its joint forest management policies in the early 1990s, gradually expanding to a national program in the mid-1990s. In India a national JFM resolution issued in 1990 provided a basis and encouragement for states to pass their own JFM policies creating local frameworks for implementation and securing financing from development agencies. While the center endorsed JFM as consistent with the forestry laws, the states held the power to define the guidelines for JFM implementation, each in their own unique way. The Philippines launched a community-based forest management policy in 1995, after a decade of experimentation with CF-styled projects and schemes and a specific program for indigenous peoples in 1993 supported by the IPRA law in 1997. In Nepal, CF became a government priority program in the early 1990s, though forestry projects had been developed in the field since the early 1980s.

New policies and legislation empowering community forest management appear to have helped leverage donor investment in this sector, even if their initial impact on securing community rights to livelihood from state forest lands has been modest. In retrospect, many of the initial policies were cautiously framed, restricting community rights to income flows from forest lands. By contrast, donor investment in CF and related participatory resource management initiatives in Asia over the past decade has totaled between two to three billion dollars, fueling a plethora of field projects, training activities, meetings and workshops, and publications, all of which have given the forest sector transition considerable energy.

Two policy strategies are emerging in Southeast Asia that support greater community involvement in forest management. The first involves the formulation and implementation of laws and policies that explicitly articulate community rights and responsibilities on lands that have historically been claimed by the state and managed by its agencies or private sector lessees. The second are policies that support devolution and decentralization, including increased local government authority over natural resource management planning, protection, and production. It is useful to compare national experiences with these two distinctly different approaches to crafting a supportive policy environment. A third strategy is the growing importance of an integrated focus on community needs through the espousing of Millennium Development Goals.[224]

Some countries have proceeded to develop specific national community forest management policies, laws, along with the administrative instruments to implement them. The Philippines and Cambodia are among such nations, though the CF laws and policies in Cambodia are still in an early phase of implementation. While CF legislation is of recent origin, the process of formulation has been unusually successful in creating a dialogue between government, donors, and the NGO community. The outcome, as described in Part III, is the creation of policy documents, sub-decrees, and implementation guidelines that are broadly acceptable to a diverse group of stakeholders. The extended period during which policy development process occurred has also created space for government line agencies and NGO field programs to disseminate information regarding the policy and prepare implementation strategies.

While the national CBFM strategy in the Philippines was approved in 1995, a variety of CF policies, laws, and programs have been emerging since the mid-1980s. Though CF policy support in the Philippines is well-articulated at many levels, from the constitution through detailed implementation orders, as a national strategy its priority has varied with administrations. Nonetheless, there has been sufficient political commitment to implement CBFM and Ancestral Domain legislation to ensure that approximately one-third of the Philippine forest lands have been formally transferred to community groups. The linkages between CF policies and decentralization policies are also creating increasing interaction between local government and communities. While policies to empower local governance have been in place for the past ten years, only recently have sub-district and district level government begun to take an active role in resource management planning activities, in part due to difficulties interacting with central government line agencies.

Thailand, Vietnam, and Indonesia are working in pre-national CF policy contexts, but have found ways to support communities in forest management by linking it to related national agendas with approved policies and legal decisions such as decentralization, commune law, community fisheries law, land law, and protected area acts. The lengthy experience of the CF bill in Thailand has often left field practitioners, within and outside government, frustrated. Prospects for approval of a meaningful bill, as noted in Part III, remain distant, in part due to the influence of middle class and urban populations' concern over strict watershed protection for their flood-prone city, and a national environmental goal of restoring forests on 40 percent of the nation's land area. National decentralization of governance functions, mandated under the new constitution, however, extend new rights to local government (*tambon*-subdistrict and *ampoe*-district) over local natural resources providing novel avenues for community leaders to inform decision making. The capacity of local government representatives to participate in management and engage with local communities requires development, as well as in interacting with the Royal Forest Department (RFD) and protected areas officials. Presently, sub-district representatives oversee fire protection and road repair, but have little involvement in production management. While there has been a growing willingness within the RFD to work with communities over the past decade, the absence of a formal CF legislation and enabling policies has constrained motivated foresters, while giving unsupportive staff an excuse to avoid CF programs. With the newly-created Ministry of Natural Resources and Environment, the RFD may be encouraged to play a larger role in community forest management.

There appears to be growing support for CF among party members and policy makers in Vietnam, in part due to the efforts of a multi-stakeholder working group to assess CF needs and prepare policy recommendations.[225] Much of the substantial progress that has been made in devolving forest protection and management to communities has been accomplished through national projects supporting household and commune forestry and through the considerable authority and freedom given to provincial and district governments to achieve broader national management objectives. There is clear recognition at the national policy level that, given the diversity of communities and practices in Vietnam, local governments will need to adapt central strategies to conform to local realities. Case studies indicate that one strategy being adopted by local governments is to recognize traditional and more recent forms of community-based forest management,[226] especially in the north, northwestern and north-central parts of the country. In northern Vietnam, local government orientation and authority is very responsive to its constituency, including ethnic minority groups. In many areas communities are encouraged to set their own rules, while local government facilitates exchanges between forest-dependent villages.

In Indonesia, CF does not yet have strong legislative backing, though there are elements in the constitution and forest laws that recognize some community rights over state forest domain. Demand for CF policies and laws have come primarily from the Indonesian NGO community and universities, and increasingly from the donor community. The Ministry of Forestry and Estate Crops is currently refocusing its social forestry program on communities by allocating certain areas for community management, but resistance to devolve control to local government and rural hamlets is apparent. The severity of illegal logging is acknowledged in the country up to the level of the presidency, including recent action by the executive branch to impose the death penalty or life imprisonment for financing illegal logging. The impact of this new regulation on rights of indigenous communities in customary forests could not be ascertained, given that indigenous rights in forest areas are still weak under prevailing Indonesian laws.[227] Local NGOs are organizing detailed documentation on violence in areas generating "conflict timber." Local governments in Indonesia, after decades of tight control by the center, are gaining authority to make a wider range of decisions, including over natural resource management, under the Law on Regional Governance, yet local authorities frequently lack the training, orientation, staff and finances, or desire to initiate support to community management.

In other countries in Asia, CF is also finding a growing role in forest management in the emerging country of Timor, where the government has already initiated a household teak planting program in some of the low-lying areas, as well as with rehabilitation of degraded forest lands through agroforestry. Government planners are well aware of the need to delineate lands and identify the role of communities in management, as community participation is viewed as a necessity given that the Department of Forest and Water Resources (DFWR) has a staff of 57 covering 1.1 million of forest lands or 58 per cent of the country's land area. The draft forest management policy incorporated a review of traditional forest management practices (*tara bandu*) and recognizes its viability as a strategy for sustainable forest management while identifying factors affecting its present effectiveness. Through the World Bank-funded Agricultural Rehabilitation Program (ARP), the Ministry facilitated the formation of forest user groups and provided them a Memorandum of Agreement as a level of tenure security while lands are being delineated.

Lao PDR, through the government's Land and Forest Allocation Program, is allocating degraded land to households for crop cultivation, tree planting and grazing. As part of the allocation process, village forests are being classified into utilization, protection and rehabilitation based on rules and agreements between the village and Provincial Agriculture and Forestry Office (PAFO). Several international NGOs are assisting government in the review of program implementation, and are assisting PAFO to address boundary issues, and other conflicts with local development projects, as well as well as the limitations in the present system for securing community tenure in forest lands.

A recent FAO review of international experiences with community forest management planning indicates that problems found in the five countries reviewed here are present in other nations as well. In Myanmar, the preparation of community forest management plans is the responsibility of the supervising forester, who may or may not consult the community.[228] The review of Myanmar doubted that CF could be effectively implemented given current levels of commercial exploitation and disempowerment and mistrust of the civilian population. In the case of Kalimantan, it has been noted that, "These simplified plans are still too complicated for CFMs to carry out by themselves, without receiving significant external professional assistance."[229] Malaysia's forest sector, with its emphasis on commercial timber harvesting, is addressing community natural resource management concerns through a national agency for tribal groups, in an effort to provide basic services and alternative livelihood opportunities to forest-dependent communities.

In reviewing some of Asia's national experiences in developing CF laws and policies, it is apparent that many countries in Asia are making progress devolving greater forest management responsibilities to communities through local government institutions and contractual agreements, while they continue to explore legal avenues to enable CF. Thailand continues to work on the passage of the Community Forestry Bill, twelve years after it initiated the first draft in 1992. Indonesia is reviewing the interface of new decentralization and forestry policies in relation to the administration of the Community Forestry Decree passed by the Ministry of Forests in 2001. Vietnam has been attempting to integrate community forest management into the country's existing legal framework since 2001. Due to the "in-process" nature of CF policy development, the CF movement has been driven as much by donor funding as by recent policies and laws, especially at the center. Local governments have another experience with CF policy implementation, much of it related to new decentralization and devolution laws and policies. Meanwhile communities are demanding greater access to basic services, better processes of consultation and transparency in decision making regarding land allocation.

In some cases, national forestry line agencies are increasingly directed to respond to the needs of local governments, shifting their accountability. In the Philippines, for example, the Nueva Vizcaya provincial government signed a memorandum of agreement with the DENR Regional Office to allow it to issue land management agreements to communities living in the Magat Forest Reserve.[230] The Wonosobo District Parliament passed a regulation that gives communities legal access to state forest lands.[231] In Vietnam, the Cao Bang provincial government used an article on village forests under the 1991 Forest Protection and Development Law to acknowledge villages organizing to manage forest lands. Many practitioners now regard community forest management as something greater than a program or project. It is increasingly seen as a nationally mandated strategy for improving forest cover and local empowerment with responsibility.

The process of formally integrating community forestry policies into the forestry sector and larger society, including creating awareness, supportive attitudes, appropriate procedures, and associated financing, is a lengthy process requiring years of pilot programs, policy development, and capacity building. Figure 4 depicts the many levels of society at which a diverse array of factors shape the outcome of forest sector transitions. National experiences from Southeast Asia indicate that this process may be viewed in phases. For example, some analysts view CFM policy implementation in the Philippines in three phases. The first phase involved a ten-year period of preliminary programs, beginning in the mid-1980s through the early 1990s. The second phase occurred after the approval of the national CBFM strategy in 1995, a period of CF activity that was heavily financed by donors including the Asian Development Bank and USAID. At present the Philippines appears to be entering a third phase of development, but this time the CBFM will probably not be supported through large development agency projects. Potentially this could have greater impact on CFM policy implementation, as mainline staff of DENR are now engaged in the process, rather than relying on contractual personnel, a major problem with CFM implementation modalities based on project financing. While the Royal Forest Department of Thailand has made considerable progress orienting staff regarding the need for CFM and implementation procedures, the integration of participatory approaches within the agency has been slowed by conflicting watershed and protected area policies. As a consequence, the mandate to move forward with CFM implementation in Thailand has been weak and diminished field level impact.

A major impediment to Asia's forest sector transition is the rigidity of the professional forestry paradigm. As a recent report noted, most professional foresters have never seriously considered indigenous forest cultures as models for management, despite their general sustainability and profitability.[232] Why? The physiognomy is so close to an old-growth or secondary forest that they are easily confused with a natural forest. The researchers suggest that it is because such forests have been invented by politically "invisible people," the same swidden farmers, who are generally considered to be bad managers of the forest. And further, that forest gardens do not "fit" the conventional separation between forests and agriculture, nor does their customary tenure system fit within the laws and regulations governing the public forest domain. Nonetheless, it is critical that the Asian forestry paradigm finds a way to integrate the concept of "forest culture on farmlands." To not do so would be to ignore "millennia of co-evolution between human societies and forests." [233]

The transition from a bureaucratic or corporate management modality of forest management to a decentralized, community-based approach is a complex and lengthy process. Furthermore, many indigenous systems of management and local community institutions have seen their authority eroded by government policies and socio-economic developments over the past century, undermining their capacity to perform past resource management functions. The engagement of communities in forest management, even in countries where enabling policies have been passed, has been resisted by elements within government agencies and in the private sector. In response, a number of organizations have emerged in Southeast Asia to provide some oversight over the implementation of CF policies and programs.

The recent formation of CF networks and advocacy bodies in Southeast Asia over the past five years has been important in establishing better linkages between isolated rural communities and the small NGOs that support them. The presence of national and provincial level CF networks and federations has also created sources of information with community perspectives that can inform development agencies and government policy makers. In some cases, development agencies are facilitating the expansion of such CF networks to help sustain their community-based resource management projects.

In the Philippines, the USAID Natural Resources Management Project assisted the national federation of the CBFM people's organizations in 1998, before the project ended. In Andhra Pradesh, IFAD and the World Bank are supporting the creation of cluster organizations and sub-district federations to reinforce the progress made in developing women-administered thrift associations. In Vietnam, a number of agencies including the EC, GTZ, IUCN, WWF, and SIDA have contributed to the operation of a national CF working group. FIPI, a national forestry research agency is facilitating the development of a provincial level network of communes and villages in the Northeast of Vietnam engaged in village forest management. In Thailand, the Northern Farmers' Network has created a Community Forestry Network, connected with the more broad-based Assembly of the Poor. In Cambodia, the national CF/CPA Network has been operating for several years, bringing together community leaders, NGOs, and forestry officials with plans to expand district level CF networks around the country.

National CF strategies are also relying on various communication media to transmit news of policies, programs, and field experiences. A Cambodian NGO, Mlup Baitong, has initiated a radio program to more widely disseminate information on community-related policies, including villager call-in talk shows. In Indonesia, a national CFM communication forum (FKKM) was established to link dispersed communities and NGOs engaged in CF activities throughout the scattered islands of the archipelago with ongoing communications based on email transmissions and annual meetings. The EU Asia Information, Technology and Communications Programme (AIT&C) is developing multi-media resource materials that feature exchanges across different Asian cultures on resource management. Networks, federations, and associations represent an important step in the evolution of the community forest management movement in Southeast Asia. Such institutions are developing the capacity to link remote rural communities into a broader civil society dialogue regarding the rights of local people to state forest resources. While these new organizations are not always able to persuade the government or donors to adopt their point of view, they are increasingly asked to participate in stakeholder planning sessions and policy reviews, a major shift from a decade earlier when policies were formed and projects designed without public consultation.

FIGURE 4: Hierarchy of Societal Contexts and Forces Influencing Community Forestry Transactions

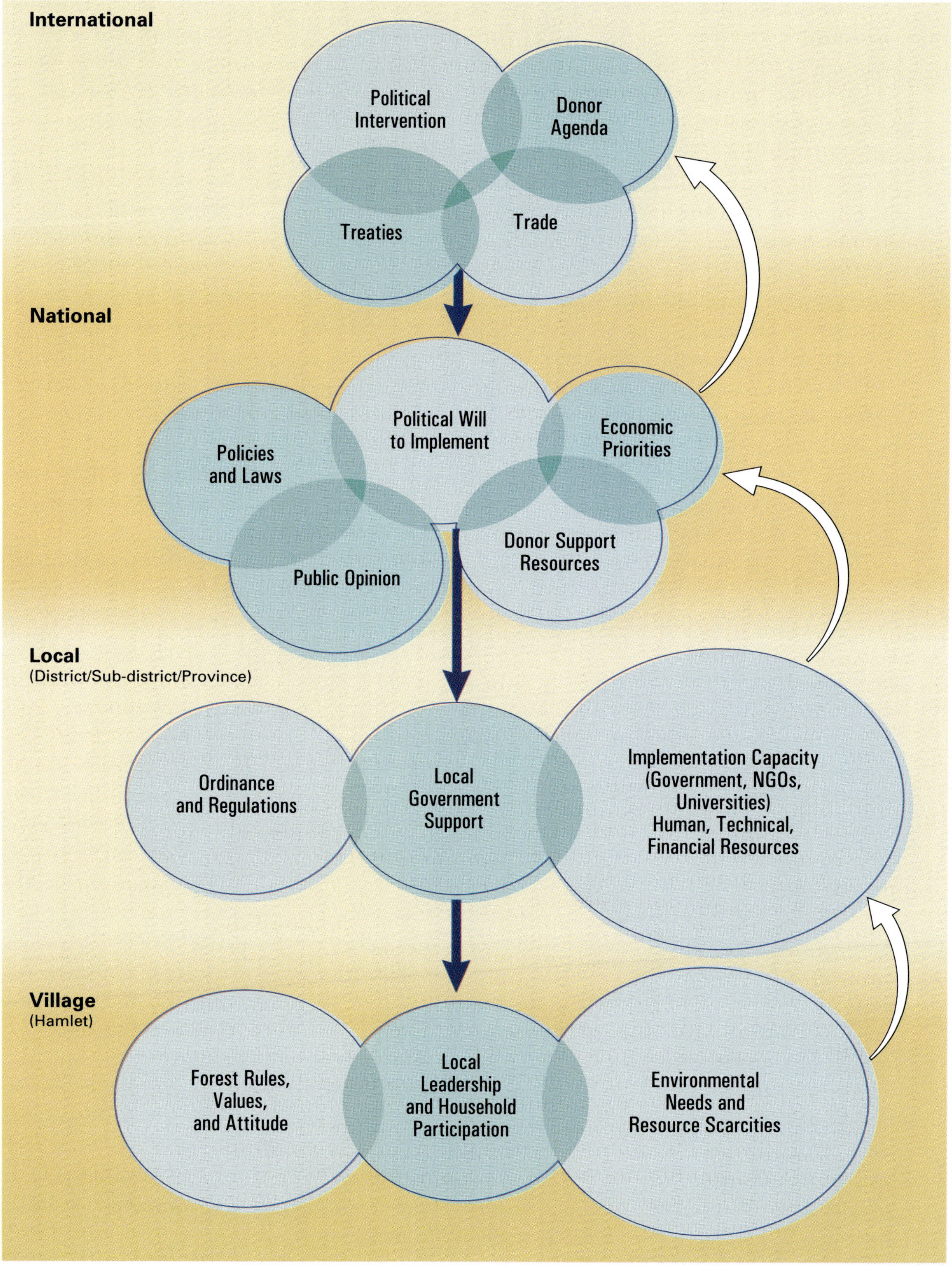

Social impact reviews and assessments are occurring in a number of CF program sites across the region, both to ensure that social equity occurs during the implementation process and to document and communicate the experience to government planners and donor agencies. In Indonesia, for example, several groups including CIFOR, ICRAF, DFID, the Community Forestry Communication Forum, and the Ford Foundation are monitoring the cases of Kutai Barat, Lampung, Malinau, and Wonosobo. In another setting, Filipino NGOs expressed their outrage in 2002 due to misuse and manipulation of the larger CBFM areas, resulting in prompt government action by the DENR. In Thailand, NGOs and academics place continuous pressure on the RFD to respond to the needs of forest-dependent communities. In Cambodia, international and local NGOs are creating greater social awareness regarding the natural resource needs and rights of rural people. The NGO Forum, a coalition of NGOs engaged in a wide spectrum of development activities in Cambodia, has formed a special group to deal with community forestry issues and has played an effective role in facilitating community inputs into new CF policy development. In Ratanakiri Province in Cambodia, an advocacy network was created specifically to address the accelerating erosion of ancestral land rights and community forestry concerns, to inform policy development, and to build local capacity to map and document communal lands. At the regional level, DFID and Ford Foundation supported a review of the effectiveness of international CF networks in achieving impact on the ground. These types of initiatives reflect the growing attention and capabilities of civil society in Southeast Asia to monitor transitions to community-based management systems and better represent program implementation experiences to governments and development agencies.

One critical area where the roles of communities and local government are changing is in the allocation and distribution of revenues from forest lands. In the Philippines, for example, the central government is gradually channeling forest resource tax revenues to local government, while in the past it was largely under the control of the DENR. In Lao PDR, the government support of CF policies, in part represents an attempt to register forest land rights of communities in order to levy land taxes and increase government income. In Indonesia, a much greater share of timber revenues will flow to the district and provincial governments from which the logs were harvested. Ultimately, policies that increase the involvement of communities in the flow of forest revenue may be one of the best indicators that a meaningful management transition is under way.

Throughout human history rural communities have acted as the stewards of the earth's forests. Over the past century, however, the role of rural people in resource management has declined as forests have been nationalized, government forest departments were established, and forest lands were leased to private companies. The erosion of indigenous cultures, their loss of authority in a modernizing world, and global commercialization have all placed growing pressure on traditions of community forest stewardship. As local institutional authority is diminished, indigenous forms of forest conservation are weakened, often leading to forest fragmentation, degradation, and loss of forest cover.

In Southeast Asia, however, indigenous people and their institutions continue to play an important role in forest protection and management. As in other parts of the world, new values emanating from modern urban cultures undermine indigenous belief systems that frequently include elements that support conservation of natural resources. The expansion of world religions and politics also introduces new forces that can erode the solidarity often found in indigenous communities. Market forces have had major impact on community forest management, especially the privatization of land resources once held under communal stewardship. The privatization of "communally" held forests and watersheds typically leads to their rapid deforestation and permanent conversion to agriculture. As the landscape of the upland regions are denuded of forest cover, much of the unique flora and fauna disappears as well. As community forests are diminished, so too are the rich and diverse cultural systems that have fostered them for generations.

With their forests and cultures endangered, rural communities across Southeast Asia are taking action. Fearing their cultures may be lost forever, and realizing the important roles that forests have played and continue to play in their existence, a growing number of communities are finding ways to strengthen their traditions while protecting their forests. Such grassroots initiatives are sprouting in many rural parts of Southeast Asia as communities recognize the growing threats to their indigenous cultures and unique environments. The revitalization of local systems of resource stewardship is a growing trend that is gaining the political, financial, and technical support of local and national government, NGOs, and development agencies. This movement provides new opportunities to reverse the deforestation and environmental damage so widespread in Southeast Asia in the Twentieth Century. This initiative depends on the people of the place, those who have a long-term vested interest in sustaining the quality of the natural environment, taking action and gradually being empowered by local and national governments. This process is ongoing not only in Southeast Asia, but around the world, in countries of the North and South. After more than a century of forest nationalization and industrial exploitation, it will take decades to re-establish more effective and sustainable systems of stewardship. The most promising process to achieve success will be strong partnerships of local communities, local government, and civil society.

References

1 Hutterer, Karl L. 1988. "The Prehistory of the Asian Rain Forests" in Julie Sloan Denslow and Christine Padoch, *People of the Tropical Rain Forest* (1988) (Berkeley, CA: University of California Press) p. 64.

2 Blackburn, Thomas C. and Kat Anderson. 1993. "Before the Wilderness: Environmental Management by Native Californians" (Menlo Park, CA: Ballena Press) p. 17.

3 Cronon, William. 1983. "Changes in the Land: Indians, Colonists, and the Ecology of New England" (New York: Hill and Wang) p. 13.

4 Diamond, Jared. 2005. "Collapse: How Societies Choose to Fail or Succeed" (New York: Viking Penguin).

5 Cronon, William. 1983. p. 13.

6 Cochester, Marcus (ed.). 2001. "A Survey of Indigenous Land Tenure" in *Forest Peoples Programme* (2001) (UK: Moreton-in-March).

7 Petherham, John and Peter Stephen and Don Gilmour. 2002. "Collaborative Forest Management – A Review" (Melbourne, Australia: University of Melbourne) p. 8.

8 Korten, David C. (ed.). 1986. "Community Management: Asian Experience and Perspectives" (West Hartford, Conn.: Kumarian Press) p. 1.

9 Parnwell, Michael J.G. and Raymond L. Bryant. 1996. "Environmental Change in Southeast Asia" (London: Routledge) Poffenberger, Mark. 1997. "Beyond Timber Barons" in *Asian Survey (Vol. XXXVII, No. 5, May 1997).*

10 Edmunds, David and Eva Wollenberg. 2003. "Local Forest Management: The Impact of Devolution Policies" (London, UK: Earthscan Publications, Ltd.).

11 Burch, William R. 1990. "Foreword" in Mark Poffenberger (ed.), *Keepers of the Forest* (1990) (West Hartford, Conn.: Kumarian Press) p. xv.

12 Korten, David C. (ed.). 1986. p. 3.

13 Korten, David C. (ed.). 1986. p. 4-5.

14 Chamber, Robert. 1983. "Rural Development: Putting the Last First" (London: Longman Group, Ltd.) p.201.

15 Salim, Emil and Ola Ullsten. 1999. "Our Forests: Our Future - Report of the World Commission on Forests and Sustainable Development" (Cambridge: Cambridge University Press) p. 9.

16 Scott, Margaret. 1989. p. 32.

17 Salim, Emil and Ola Ullsten. 1999. p. 58.

18 Akiko Inoguchi, Rowena Soriaga and Peter Walpole. 2005. "Approaches to Controlling Illegal Forest Activities: Considerations from Southeast Asia (Working Paper Series No. 7)" (Bohol, Philippines:Asia Forest Network).

19 United Nations. 2000. "2000 Millenium Development Goals" (United Nations).

20 Poffenberger, Mark (ed.). 1990. "Keepers of the Forest: Land Management Alternatives in Southeast Asia" (West Hartford, Conn.: Kumarian Press).

21 Hill, Ronald. 2002. "Southeast Asia: People, Land and Economy" (Canberra: Allend and Unwin).

22 Reid, Anthony. 1998. "Humans and Forests in Pre-Colonial Southeast Asia" in Richard Grove (ed.), *Nature and the Orient: The Environmental History of South and Southeast Asia (1998)* (Delhi: Oxford University Press) p. 112.

23 Reid, Anthony. 1998. p. 113.

24 Reid, Anthony. 1998. p. 116.

25 Bruce, John W. and L. Fortmann. 1992. "Property and Forestry" in P.N. Nemetz (ed.), *Emerging Issues in Forest Policy (1986)* (Vancouver: University of British Columbia Press) p. 4.

26 Bruce, John W. 1999. "Legal Bases for the Management of Forest Resources as Common Property (Community Forestry Note 1)" (Rome: FAO) p. 6.

27 Bruce, John W. 1999. p. 7.

28 Poffenberger, Mark (ed.). 1990.

29 Bruce, John W. 1999. p. 8.

30 Bromley, Daniel W and Michael M. Cernea. 1989. "The Management of Common Property Natural Resources (World Bank Discussion Paper No. 57)" (Washington, D.C.: The World Bank) p. iii.

31 Poffenberger, Mark (ed.). 1990. p. xix.

32 Poffenberger, Mark (ed.). 1996. "Communities and Forest Management" (Cambridge, UK: IUCN) p. 1.

33 Walpole, Peter. 2002. "An Analysis of Drivers and Impacts of Landuse Change in the Tropical Uplands of Mindanao, Philippines (dissertation)" (London: King"s College).

34 Fox, Jefferson. (unpublished). "Mapping a Changing Landscape: Land Use, Land Cover, and Resource Tenure in Northeastern Cambodia."

35 Primack, Richard B. and Thomas E. Lovejoy. 1995. "Ecology, Conservation, and Management of Southeast Asian Rainforests" (New Haven: Yale University Press) p. 11.

36 McDowell, M.A. "Development and the Environment in ASEAN (Pacific Affairs, 62)," p. 307-329.

37 Poffenberger, Mark. 1997. "Rethinking Indonesian Forest Policy: Beyond the Timber Barons" in *Asian Survey (Vol. XXXVII, No. 5, May 1997)*. p. 466.

38 Richardson, Michael. 2002. "Donors Seek Curbs on Logging, But Powerful Interests are Involved" in *International Herald Tribune (February 1, 2002).*

39 Akiko Inoguchi, Rowena Soriaga and Peter Walpole. 2005.

40 Jakarta Post. 1994. in *Jakarta Post (October 3, 1994).*

41 FAO. 2001. in *State of the World's Forest: 2001 (2001)* (Rome: FAO) p. 155.

42 Barber, C.V. and J. Schweithelm. 2000. "Trial by Fire: Forest Fires and Forestry Policy in Indonesia's Era of Crisis and Reform" (Washington, D.C.: World Resources Institute).

43 Waggener, T.R. 2001. "Logging Bans in Asia and the Pacific: An Overview" in Patrick Durst et al (ed.) *Forest Out of Bounds: Impacts and Effectiveness of Logging Bans in Natural Forests in Asia-Pacific.* (Bangkok, Thailand: FAO-RAP) p. 1-6.

44 Forest Out of Bounds: Impacts and Effectiveness of Logging Bans in Natural Forests in Asia-Pacific. (Bangkok, Thailand: FAO-RAP).

45 Ribot, Jesse C. 2002. "Democratic Decentralization of Natural Resources" p. 4.

46 Ribot, Jesse C. 2002. p. 5.

47 Dupar, Mairi and Nathan Badenoch. 2002. "Environment, Livelihoods, and Local Institutions: Decentralization in Mainland Southeast Asia" (Washington, D.C.: World Resources Institute) p. 46.

48 Sayer, Jeffrey. 2002. "Foreword" in Carol J. P. Colfer and Ida Aju Pradnja Resosudarmo (eds.), *Which Way Forward? People, Forests, and Policy-Making in Indonesia (2002)* (Washington, D.C.: Resources for the Future) p. xvii.

49 Bruce, John W. 1999. p. 21.

50 Bruce, John W. and L. Fortmann. 1992. p. 471-496.

51 Ostrom, E. 1986. "Issues of Definition and Theory: Some Conclusions and Hypothesis (from Proceedings of the Conference on Common Property Resource Management, April 1985, Washington D.C., cited By J.W. Bruce and L. Fortmann - 1992)" in P.N. Nemetz (ed.), *Emerging Issues in Forest Policy (1986)* (Vancouver: University of British Columbia Press) p. 471-496.

52 Bruce, John W. 1999. p. 6-7.

53 Bruce, John W. 1999. p. 10.

54 Bruce, John W. 1999. p. 12.

55 Poffenberger, Mark and Chatrapati Singh. 1995. "Communities and the State: Re-Establishing the Balance in Indian Forest Policy" in Mark Poffenberger and Betsy McGean (eds.), *Keepers of the Forest (1995)* (New Delhi: Oxford University Press) p. 71.

56 Bruce, John W. 1999. p. 21.

57 Bruce, John W. 1999. p. 21.

58 Bruce, John W. 1999. p. 1.

59 Bruce, John W. 1999. p. 7.

60 Edmunds, David and Eva Wollenberg. 2003. p. 175.

61 Independent Forest Sector Review. 2004. "The Forest Sector in Cambodia, Part I: Policy Choices, Issues, and Options" (Phnom Penh: Independent Forest Sector Review).

62 FAO. 2001.

63 Forest Trends. 2002. in *Who Owns the World's Forests?: Forest Tenure and Public Forests in Transition (2002)* (Washington, D.C.: Forest Trends).

64 Pingle, Urmila. 2004. "Communities & Forest Management: Janaram Forest Divison, Adilabad, Andhra Pradesh, India" in *Community Forestry International, December 2004* (Santa Barbara, CA) p. 1.

65 Pingle, Urmila. 2004. p. 2.

66 ESSC. 2004. (Bohol: Environmental Science for Social Change).

67 McKenney, Bruce. 2002. "Questioning Sustainable Concession Forestry in Cambodia" in *Cambodia Development Review (Vol. 6, Issue 1, January-March 2002),* p. 1.

68 The World Bank. 1996. "Forest Policy Assessment: Cambodia" (Washington, D.C.: The World Bank) p. 1.

69 DAI. 1998. "Findings and Recommendations of the Log Monitoring and Log Control Project: Main Report" (Betshseda, MD: Development Alternatives, Inc.).

70 ARD. 2002. "Proceeds of Meeting to Brief the International Community and Senior Officials of the Royal Government of Cambodia on Forest Policy Reform (cited by Bruce Mckenney - 2002)" in *Cambodia Development Review (Vol. 6, Issue 1, January-March 2002).*

71 Fraser, Thomas. 2000. "Cambodia Forest Concessionaire Review Report" (Phnom Penh: Asian Development Bank).

72 McKenney, Bruce. 2002.

73 KPGM. 2001. "The Equitability of the Forest Taxation System in Cambodia (cited by Bruce McKenney - 2002)" in *Cambodia Development Review (Vol. 6, Issue 1, January-March 2002)* p. 3.

74 McGrath, Bill. 2003. "(personal communication)."

75 McKenney, Bruce. 2002. p. 5-6.

76 Evans, Patrick and Melissa Marschke and Kiran Paudyal. 2004. "Flood Forests, Fish, and Fishing Villages: Tonle Sap, Cambodia" in *Asia Forest Network, 2004(Bohol, Philippines)* p. 10.

77 Royal Government of Cambodia. 2002. "Final Draft National Community Forestry Strategic Plan" (Phnom Penh: Department of Forests and Wildlife).

78 RECOFTC. 2002. "Balancing Ownerhsip: Overview of Writing Retreat about the Community Forestry Consultation Process in Cambodia, Phuket, Thailand" (Thailand: RECOFTC).

79 Independent Forest Sector Review. 2004.

80 Evans, Tom D. et al. 2003. "A Study of Resin-Tapping and Livelihoods in Southern Mondulkiri Cambodia, with Implications for Conservation and Forest Management" (Phnom Penh: World Conservation Society).

81 Fichtenau, Jurgen et al. 2002. "An Assessment of Ongoing Community Forestry Initiatives in Cambodia" (Phnom Penh: Cambodian-German Forestry Project).

82 Fichtenau, Jurgen et al. 2002.

83 Bradley, Amanda and Mark Poffenberger and Ponreay. 2005. "National Survey of Community Forestry in Cambodia" in *Community Forestry Alliance for Cambodia* (draft) (Phnom Penh, Cambodia).

84 Brown, Graeme and Alistair Stephens. 2005. "Mapping of Possible Community Forest Management Tenures in Ratanakiri Province, Cambodia" in *Community Forestry Alliance for Cambodia (CFAC) (* forthcoming) (Phnom Penh, Cambodia) p. 26.

85 Osborne, Milton. 2000. "The Mekong: Turbulent Past, Uncertain Future" (New York: Grove Press) p. 231-260.

86 Agence France-Presse. 2003. "Cambodian Prime Minister Voices Regret over Forest Losses" (Agence France-Presse).

87 Independent Forest Sector Review. 2004. p. 68.

88 Wrangham, Rachel. 2002. "Changing Policy Discourses and Traditional Communities, 1960-1999" in Carol J. P. Colfer and Ida Aju Pradnja Resosudarmo (eds.), *Which Way Forward? People, Forests ,and Policy Making in Indonesia (2002)* (Washington, D.C.: Resources for the Future) p. 20-22.

89 Peluso, Nancy Lee. 1990. "A History of State Forest Management in Java" in Mark Poffenberger (ed.), *Keepers of the Forest (1990)* (West Hartford, Conn.: Kumarian Press) p. 42-43.

90 Wrangham, Rachel. 2002. p. 23.

91 Wrangham, Rachel. 2002. p. 24.

92 Lindayati, Rita. 2002. "Ideas and Institutions in Social Forestry Policy" in Carol J. P. Colfer and Ida Aju Pradnja Resosudarmo (eds.), *Which Way Forward? People, Forests, and Policy Making in Indonesia (2002)* (Washington D.C.: Resources for the Future) p. 45.

93 Lindayati, Rita. 2002. p. 40.

94 Lindayati, Rita. 2002. p. 52.

95 Lindayati, Rita. 2002. p. 52.

96 Lindayati, Rita. 2002. p. 52.

97 Fay, Chip and Martua Sirait. 2002. "Reforming the Reformists in Post-Soeharto Indonesia" in Carol J. P. Colfer and Ida Aju Pradnja Resosudarmo (eds.), *Which Way Forward? People, Forests, and Policy Making in Indonesia (2002)* (Washington D.C.: Resources for the Future) p. 141.

98 Fay, Chip and Martua Sirait. 2002. p. 142.

99 Ridwan, Agus. 2003. "(personal communication)."

100 District of Wonosobo Government. 2001. "District Regulation of Wonosobo: Number 22, Year 2001, Regarding Community-Based Forest Resources Management."

101 World Bank, Indonesia. 1989. "Forest, Land and Water: Issues in Sustainable Development (Report No. 7822-Ind)" (Washington D.C.: World Bank Asia Regional Office).

102 FEER. 1990. "Forestry: A Saw Point for Ecology" in *Far Eastern Economic Review (No. 19, April 1990)* (Far Eastern Economic Review) p. 60 Dauvergne, Peter. 2001. "Loggers and Degradation in the Asia-Pacific" (Cambridge, UK: Cambridge University Press) p.16-17.

103 Colfer, Carol J. P. and Ida Aju Pradnja Resosudarmo (eds.). 2002. "Which Way Forward? People, Forests, and Policy Making in Indonesia (2002)" (Washington D.C.: Resources for the Future) p. 3.

104 Dermawan, Ahmad and Ida Aju Pradnja Resosudarmo (eds.). 2002. "Forests and Regional Autonomy: The Challenge of Sharing the Profits and Pains" in Carol J. P. Colfer and Ida Aju Pradnja Resosudarmo (eds.), *Which Way Forward? People, Forests, and Policy Making in Indonesia (2002)* (Washington D.C.: Resources for the Future) p. 340.

105 Dermawan, Ahmad and Ida Aju Pradnja Resosudarmo (eds.). 2002.

106 Poffenberger, Mark. 1999. "Communities and Forest Management in Southeast Asia" in Mark Poffenberger (ed.), *Keepers of the Forest (1990)* (Berkeley, CA: IUCN).

107 Alcorn, Janis. B. and Antoinette G. Royo (eds.). 2000. "Indigenous Social Movements and Ecological Resiliences: Lessons from the Dayak of Indonesia" (Washington D.C.: Biodiversity Support Project) p. 3.

108 Poffenberger, Mark. 1990. "Facilitating Change in Forestry Bureaucracies" (West Hartford, Conn.: Kumarian Press) p. 101.

109 Alcorn, Janis. B. and Antoinette G. Royo (eds.). 2000.

110 FWI/GFW. 2002. "The State of the Forest: Indonesia" (Bogor, Indonesia: Forest Watch Indonesia and Global Forest Watch) p. 13.

111 Arupa, Yayasan Koling, and the Asia Forest Network. 2004. "Communities Transforming Forestland: Java, Indonesia" (Bohol, Philippines: Asia Forest Network).

112 ESSC. 1999. Philippine Culture and Ecosystems (poster map).

113 Poffenberger, Mark. 1990. in Mark Poffenberger (ed.), *Keepers of the Forest (1990)* (West Hartford, Conn.: Kumarian Press) p. 12.

114 Lynch, Owen. (unpublished). "The Ancestral Land Rights of Tribal Indonesians: A Comparison with the Philippines" p. 6.

115 Pulhin, Juan and Perlyn Pulhin. 2003. "Community–Based Forest Management in the Philippines: Retrospect and Prospects" in M.Inoue and H. Isozaki (eds.), *People and Forest – Policy and Local Reality in Southeast Asia, the Russian Far East and Japan (2003)* (Dordrecht: Kluwer Academic Publishers).

116 Roth, Dennis. 1983. "Philippine Forests and Forestry: 1565-1920" in Richard Tucker and J.F. Richards (eds.), *Global Deforestation and the Nineteenth Century World Economy (1983)* (Durham, NC: Duke Press Policy Studies) p. 44.

117 ADB. 1994. "ADB Approves New Forestry Policy for Asia-Pacific Region" (Manila: Asian Development Bank).

118 Poffenberger, Mark and Betsy McGean (eds.). 1993. "Upland Philippine Communities: Guardians of the Final Forest Frontier (Research Network Report No. 4)" (Berkeley, CA: Asia Forest Network) p. 61.

119 DENR-FMB. 1999. "Reforestation Statistics for 1998" (Quezon City, Philippines: Department of Environment and Natural Resources, Philippines, and Asian Development Bank).

120 ADB. 1998. "ADB Annual Report 1998" (Manila: Asian Development Bank).

121 Nissen, Todd M. and Midmore, David J. 2001. "Can Smallholder Farmers Realistically Contribute to National Reforestation Efforts in the Philippines?" in *http://www.aae.wisc.edu/sanrem-sea/2001conference/papers/paper-nissen-doc.*

122 Walpole, Peter. 2002.

123 Andin, Neria A. 2002. "Forestry Programs and Policies in Support to Tree Farming (paper presented at the 2nd National Tree Farmers Congress, Butuan City, 11-13 July 2002)" Baker, Terry. 2002. "New Zealand Experience in Tree Farming: Ambassador's Address to the 2nd National Tree Farming Congress, 12 July 2002."

124 Vitug, Marites. 2000. "Forest Policy and National Politics" in Peter Utting (ed.), *Forest Policy and Politics in the Philippines: The Dynamics of Participatory Conservation (2002)* (Manila: Ateneo de Manila University Press) p. 15-20.

125 Ross, Michael. 1996. "Conditionality and Forestry Reform in the Philippines and Indonesia". in Institutions for Environmental Aid: Pitfalls and Promise" in Peter Utting (ed.), *Forest Policy and Politics in the Philippines: The Dynamics of Participatory Conservation (2002)* (Manila: Ateneo de Manila University Press).

126 Vitug, Marites. 2000. p. 22.

127 Vitug, Marites. 2000. p. 22.

128 Vitug, Marites. 2000. p. 24.

129 Vitug, Marites. 2000. p. 24.

130 Vitug, Marites. 2000. p. 25.

131 Miclat, Sylvia. 2002. "Community Forest Management in the Philippines Circa 2002 (presented at the Second AFN Regional Policy Workshop, Siem Reap, October 2002)."

132 Walpole, Peter. 2002.

133 Lindsay, Jon. 2005. "(personal communication)" (Rome, FA).

134 DENR-FMB. 2003. "The Revised Master Plan for Forestry Development (draft report)" (Manila: Department of Environment and Natural Resources, Philippines) p. 36.

135 Vitug, Marites. 1993. "Power from the Forest: The Politics of Logging" (Philippine Center for Investigative Journalism) p. 33.

136 ESSC and AFN. 2004. "The State of Community Forest Management in Bohol Province, Philippines -2004 (internal review paper)" (Bohol, Philippines: Asia Forest Network).

137 Pragtong, Komon and David E. Thomas. 1990. "Evolving Management Systems in Thailand" in Mark Poffenberger (ed.), *Keepers of the Forest (1990)* (West Hartford, Conn.: Kumarian Press) p. 167- 170.

138 Pragtong, Komon and David E. Thomas. 1990. p. 171.

139 Poffenberger, Mark (ed.). 1999. "Communities and Forest Management in Southeast Asia" (IUCN) p. 59.

140 Department of Forest Management. 1985. "Land Use Resource Management in North-East Thailand" (Bangkok: Kasetsart University) p. 132.

141 Pragtong, Komon. 1998. "Recent Decentralization Plans of the Royal Forest Department and their Implications for Management in Thailand (International Seminar on Decentralization and Devolution of Forest Management in Asia and the Pacific, Davao, Philippines, December 1998)."

142 Pragtong, Komon. 1998.

143 Mairi, Dupar and Nathan Badenoch. 2002. "Environment, Livelihood, and Local Institutions: Decentralization in Mainland Southeast Asia" (Washington D.C.: World Resources Institute) p. 12-13.

144 Royal Forest Department. 1992. "Thai Forestry Sector Master Plan" (Bangkok: Royal Forest Department).

145 Makarabhirom, Pearmsak. "(personal communication: History of Community Forestry Bill Evolution)."

146 Janchitfah, Supara. 2002. "(Report: March 24, 2002)" (Bangkok: Bangkok Post).

147 Hongthong, Pennapa. 2002. "(Report: March 27, 2002)" (The Nation).

148 Hongthong, Pennapa. 2002.

149 Hongthong, Pennapa. 2002.

150 Janchitfah, Supara. 2002.

151 Hongthong, Pennapa. 2002.

152 Wongruang, Piyaporn. 2005. "(Report: February 8, 2005)" (Bangkok: Bangkok Post).

153 Kunjit, Chaleo. 2003. "(personal communication)."

154 Poffenberger, Mark (ed.). 1999. p. 54.

155 Poffenberger, Mark and Betsy McGean (eds.). 1993. "Community Allies: Forest-Co-Management in Thailand" (Berkeley, CA: Asia Forest Network) p. 6.

156 Rojanapaiwong, Sukran (ed.). 2000. "State of the Thai Environment 1997-1998" (Greenworld Foundation) p. 79.

157 Pragtong, Komon and David E. Thomas. 1990. p. 175.

158 Odechao, Jorni. 1999. "(personal interview)" in Mark Poffenberger (ed.), *Communities and Forest Management in Southeast Asia (1999)* (IUCN) p. 102.

159 Ivarsson, Soren. 2001. "Man, Nature and Environmentalism in Thailand: The Role of Buddhism" in Ewe Poulsen, et al. (eds.), Forest in Culture - Culture in Forest, Perspectives from Northern Thailand (2001) (Denmark: Research Centre on Forest and People in Thailand) p. 43.

160 Ganjanapan, Anan. 2000. "Local Control of Land and Forest: Cultural Dimensions of Resource Management in Northern Thailand (Monograph Series No. 1)" (Chiang Mai, Thailand: RCSD) p. 162-163.

161 World Bank, Vietnam. 1995. "Environmental Program and Policy Priorities for a Socialist Economy in Transition" (Washington, D.C.: World Bank).

162 Ministry of Forestry, Vietnam. 1991. "Thirty Years of Construction and Development of Forestry 1961-1990" (Hanoi: Statistical Publishing House).

163 Le Trong Cuc, A.T. Rambo, Keith Fahrney, Tran Duc Vien, Jeff Romm, and Dang Thi Sy (eds.). 1996. "Red Books, Green Hills: Economic Reform and Restoration Ecology in the Midlands of Northern Vietnam" (Honolulu: East-West Center).

164 Poffenberger, Mark. 1999. p. 59.

165 Pham Xuan Phuong. 2001. "An Overview of the Forestry Policies Framework Related to Community Forest Management in Vietnam" in *National Workshop on a Policy Framework to Support Community*

Forest Management in Vietnam (Hanoi: National Working Group on Community Forest Management in Vietnam) p. 1.

166 Pham Xuan Phuong. 2001. p. 1-8.

167 Minh, Vu Hoa and Hans Warfinge. 2002. "Issues in Management of Natural Forests by Households and Local Communities of Three Provinces in Vietnam: Hoa Binh, Nghe An, Thua Thien-Hue (Working Paper Series Vol. 5)" (Asia Forest Network).

168 Minh, Vu Hoa and Hans Warfinge. 2002. p. 23-24.

169 Minh, Vu Hoa and Hans Warfinge. 2002. p. 32-33.

170 Minh, Vu Hoa and Hans Warfinge. 2002. p. 45.

171 Minh, Vu Hoa and Hans Warfinge. 2002. p. 45-46.

172 Nguyen Huy Dzung, Ho Manh Tuong, R. Soriaga, P. Walpole. 2004. "The Return of Limestone Forests, Northeastern Vietnam" (Philippines: Asia Forest Network).

173 Poffenberger, Mark (ed.). 1998. "Stewards of Vietnam's Upland Forests (Research Network Report No. 10)" (Berkeley, CA: Asia Forest Network) p. 23.

174 Poffenberger, Mark (ed.). 1998. p. 9.

175 Poffenberger, Mark (ed.). 1998.

176 National Working Group on CFM. 2001. "National Workshop on a Policy Framework to Support Community Forest Management in Vietnam" p. 3.

177 Sikor, Thomas. 1998. "Forest Policy Reform: From State to Household Forestry" in Mark Poffenberger (ed.), *Stewards of Vietnam's Upland Forests (Research Network Report No. 10, 1998)* (Berkeley, CA: Asia Forest Network) p. 26.

178 Thang, Nguyen Van. 1995. "The Hmong and Dzao Peoples in Vietnam" in Terry Rambo, et al. (eds.), *The Challenges of Highland Development in Vietnam* (Honolulu: East-West Center) p. 15.

179 Vu Van Me. 2003. "(presentation)."

180 CIFOR, et al. 1999. "Creating Space for Local Forest Management (In a Study of Three Nations Funded by IFAD, a Multi-Institutional Team Found that Many Policies Failed to Provide Sufficient Resource Rights to Allow Communities to Effectively Manage or Profit thereby from Decentralization Policies.)" (CIFOR).

181 Lindsay, J. 1998. "Designing Legal Space: Law as an Enabling Tool in Community-Based Management (International CBNRM Workshop, Washington D.C., May 10-14, 1998)" in Hiroyuki Tanaka, *Forest Management Plans for Collective Forest Managers in Developing Countries: A Review of Constraints and Promising Experiences (2002)* (Rome, FA) p. 73.

182 Tanaka, Hiroyuki. 2002. "Forest Management Plans for Collective Forest Managers in Developing Countries: A Review of Constraints and Promising Experiences" (Rome, FA) p. 73.

183 De Beer, Jenne and Melanie J. McDermott. 1989. "The Economic Value of Non-Timber Forest Products in Southeast Asia" (Glan, Switzerland: IUCN) p. 71.

184 European Commission. 2002. "Strategic Guidelines and Priorities for Interventions to be Financed in 2002 and 2003 under Regulation (Ec) No. 2494/2000 of the European Parliament and of the Council on Measures to Promote the Conservation and Sustainable Management of Tropical Forests and other Forests in Developing Countries (Dev/2792/2002)" (European Commission).

185 European Commission. "http://europa.eu.int/comm/internal_market/publicprocurement/keydocs_en.htm" (European Commission).

186 DFID. "(DFID Website)" (DFID).

187 Kishor, Nalin. 2004. "(personal communication)."

188 World Bank. "Environmental Strategy Paper" p. 89.

189 Fraser, Alastair. 2000. "Regional Study on Forest Policy and Institutional Reforms: Analysis and Review of ADB's Portfolio and 1994 Policy on Forestry" (Asian Development Bank).

190 ADB. 2003. "Forest Policy (working paper)" (Asian Development Bank).

191 UN. "Proposals for Action of the Intergovernmental Forum on Forests (http://www.un.org/esa/forests/docs/IFF%20PFA%20from%20IFF4%20Report.doc)" (United Nations).

192 Dempsey, Jessica. 2002. "Community Forests: Emancipatory Change or Smoky Mirrors" in *World Rainforest Movement, Bulletin No. 63, Oct. 2002* (World Rainforest Movement).

193 World Rainforest Movement. 2002. "Our Viewpoint: Community Forest Management: A Feasible and Necessary Alternative" in *World Rainforest Movement, Bulletin No. 61, Aug. 2002* (World Rainforest Movement).

194 Samabuddhi, Kultida. 2002. "Chief under Review, Probe of Auction" in *Bangkok Post, November 24, 2002* (Bangkok Post).

195 Samabuddhi, Kultida. 2002.

196 Samabuddhi, Kultida. 2002.

197 Tanaka, Hiroyuki. 2002. p. 28.

198 Donnelly-Roark, P. K. Ouedraogo and X. Ye. 2001. "Can Local Institutions Reduce Poverty in Burkina Faso?" in Hiroyuki Tanaka, *Forest Management Plans for Collective Forest Managers in Developing Countries: A Review of Constraints and Promising Experiences (2002)* (Rome, FA) p. 27.

199 Poffenberger, Mark. 1995. "Grassroots Forest Management: The Emergence of Forest Protection in Eastern India" (Berkeley, CA: Asia Forest Network).

200 Edmunds, David and Eva Wollenberg. 2001. "Historical Perspectives on Forest Policy Change in Asia - An Introduction" in Hiroyuki Tanaka, *Forest Management Plans for Collective Forest Managers in Developing Countries: A Review of Constraints and Promising Experiences (2002)* (Rome, FA) p. 52.

201 Tanaka, Hiroyuki. 2002. p. 64.

202 Edmunds, David and Eva Wollenberg. 2001.

203 Tanaka, Hiroyuki. 2002. p. 65.

204 Edmunds, David and Eva Wollenberg. 2003. p. 158.

205 Edmunds, David and Eva Wollenberg. 2003. p. 159.

206 Edmunds, David and Eva Wollenberg. 2003. p. 159-162.

207 Poffenberger, Mark (ed.). 1999.

208 Kunjit, Chaleo. 2003.

209 Asia-Pacific Forestry Commission. 2001. "Forest Out of Bounds: Impacts and Effectiveness of Logging Bans in Natural Forests in Asia-Pacific" (FAO Regional Office for Asia and Pacific)

210 Ministry of Forestry, Vietnam. 1998. "Asia-Pacific Forestry Sector Outlook Study: Country Report – Indonesia (www.fao.org)" (FAO-Rome and FAO-RAP).

211 McKenney, Bruce. 2001. "Baseline Assessment of Cambodia's Forestry Sector to Support a Feasibility Study of Forest Management Certification" (WWF) p. 15.

212 ESSC. 2003. Poster Map on the Formulation of a Community Resource Management Plan using Community Resource Mapping and Accounting (CRMA), Geographic Information Systems (GIS) and Remote Sensing (RS) in Maasin Watershed, Iloilo (Tagbilaran, Philippines). .

213 de Jong, W., M. van Noordwijk, Martua Sirait, N. Liswanti, and Suyanto. 2001. "Farming Secondary Forests in Indonesia" in *Journal of Tropical Forest Science No. 13 (2001)* p. 705-726.

214 van Noordwijk, M., James M. Roshetko, Murniati, Marian de los Angeles, Suyanto, Chip Fay and Thomas P. Tomich. 2003. "Agroforestry is a Form of Sustainable Forest Management: Lessons from South East Asia (paper presented at the UNFF Intersessional Experts Meeting on the role of Planted Forests in Sustainable Forest Management Conference, 24-28 March 2003, Wellington, New Zealand)."

215 van Noordwijk, M., et al. 2003.

216 Fay, Chip. 2002. "(paper with information based on Pusat Perpetaan Hutan, Badan Planologi Kehutanan)."

217 Fay, Chip. 2002. "(personal communication - with information based on Harry Santosa)."

218 Michon, G. 2005. "Domesticating Forests: How Farmers Manage Forest Resources" in *FORRESASIA* (Bogor, Indonesia).

219 Michon, G. and H. de Foresta. 1995. "The Indonesian Agro-Forest Model" in P. Halladay and D.A. Gilmour (eds.), *Conserving Biodiversity Outside Protected Areas: The Role of Traditional Ecosystems (1995)* (Glan, Switzerland: IUCN).

220 Edmunds, David and Eva Wollenberg. 2003. p. 150.

221 Edmunds, David and Eva Wollenberg. 2003. p. 158.

222 Edmunds, David and Eva Wollenberg. 2003. p. 158-163.

223 Edmunds, David and Eva Wollenberg. 2003. p. 73.

224 United Nations.

225 MARD. 1991. "Report by the National Working Group on Community Forest Management" (Ministy of Agriculture and Rural Development).

226 Poffenberger, Mark (ed.). 1998. Minh, Vu Hoa and Hans Warfinge. 2002.

227 FERN. 2004. "Current Regulation Poses Serious Threat" in *EU Forest Watch, September 2004* (EU).

228 Tanaka, Hiroyuki. 2002. p. 27.

229 Tanaka, Hiroyuki. 2002. p. 27.

230 AFN. 2002. "Regional Workshop on Strategies for CFM Diagnosis and Planning" (Philippines: Asia Forest Network).

231 Arupa, Yayasan Koling, and the Asia Forest Network. 2004.

232 Michon, G. 2005.

233 Michon, G. 2005. p. 7.

Annex

ANNEX 1. Equivalent Terms for Units of Governance in Southeast Asia

	CAMBODIA	INDONESIA	PHILIPPINES	THAILAND	VIETNAM
Hamlet (no legal status)	Phum–Thmei *(20-50 hh)*	Dusun *(50+ hh)*	Purok *(20-50 hh)*	Klum Ban/Pok *(20-30 hh)*	Cum *(20-30 hh)*
Village	Phum *(100+ hh)*	Desa *(5-10 dusun)*	Barangay *(7-10 purok)*	Moo Ban *(50+ hh)*	Bàn/Lang *(30-60 hh)*
Commune (local government unit)	Khum *(10-20 phum)*			Tambon *(8-15 moo ban)*	Xa *(8-15 bàn)*
Sub-district		Kecamatan *(10-15 desa)*			
District *(8+ Khum)*	Srok *(10+ kecamatan)*	Kapubaten or Kota *(10+ barangay)*	Munisipyo *(8-10 tambon)*	Amphoe *(12-18 xa)*	Huyên
Provincial	Khaet *(5+ Srok)*	Propinsi *(5-40 kabupaten)*	Probinsiya *(10+ munisipyo)*	Chang Wat *(8-20 amphoe)*	Tinh *(8-16 Huyên)*
National Government	Kingdom of Cambodia *(23 provinces)*	Pemerintah Pusat Indonesia *(32 provinces)*	Republic of the Philippines *(79 provinces)*	Royal Thai Government *(76 provinces)*	Socialist Republic of Vietnam *(62 provinces)*
Population	10 million	207 million	81 million	66 million	78 million

Index